T0388975

International Perspectives on Aging

Volume 26

The study of aging is continuing to increase rapidly across multiple disciplines. This wide-ranging series on International Perspectives on Aging provides readers with much-needed comprehensive texts and critical perspectives on the latest research, policy, and practical developments. Both aging and globalization have become a reality of our times, yet a systematic effort of a global magnitude to address aging is yet to be seen. The series bridges the gaps in the literature and provides cutting-edge debate on new and traditional areas of comparative aging, all from an international perspective. More specifically, this book series on International Perspectives on Aging puts the spotlight on international and comparative studies of aging.

More information about this series at http://www.springer.com/series/8818

Patrick L. Hill • Mathias Allemand
Editors

Personality and Healthy Aging in Adulthood

New Directions and Techniques

Editors
Patrick L. Hill
Department of Psychological & Brain Sciences
Washington University in St. Louis
Saint Louis, MO, USA

Mathias Allemand
Department of Psychology and University Research Priority Program Dynamics of Healthy Aging
University of Zurich
Zürich, Switzerland

ISSN 2197-5841 ISSN 2197-585X (electronic)
International Perspectives on Aging
ISBN 978-3-030-32055-3 ISBN 978-3-030-32053-9 (eBook)
https://doi.org/10.1007/978-3-030-32053-9

This Springer imprint is published by the registered company Springer Nature Switzerland AG.
The registered company address is: Gewerbestrasse 11, 6330 Cham, Switzerland

Acknowledgments

The current volume represents the product of a workshop on Personality and Healthy Aging hosted at the University of Zurich, due to the generous funding from the Swiss National Science Foundation.

Contents

Chapter 1
An Introductory Overview on Personality and Healthy Aging: Setting a Foundation for the Current Volume

Patrick L. Hill and Mathias Allemand

The importance of promoting healthy aging has never been clearer. Individuals are living longer lives than ever before, which places greater importance on identifying factors that promote health maintenance and improve quality of life. Though the final endpoint may be the same for all, individuals differ greatly in the extent to which they enact healthy lifestyle behaviors across the lifespan, and in their likelihood for experiencing negative health risks. In recent decades, efforts to identify why some individuals experience more or less positive aging trajectories have pointed to the value of considering personality science. Though the notion that individuals' personality dispositions are valuable for predicting health outcomes is not new (for a review of the classic work on trait anger and hostility, see Siegman, 1994), what has changed in recent years is (a) our definitions, taxonomies, and understanding of personality dispositions, (b) the knowledge base regarding why and for whom personality characteristics lead to healthier aging outcomes, and (c) the methodological and analytic approaches taken for studies in this field.

The current volume reflects an effort to present new findings, developments, and techniques in order to continue progress for research on personality traits and healthy aging. The focus is less on absolute coverage of any one domain of research or methodological expertise, and instead is intended to provide a smattering of new ideas and theoretical insights from some of the researchers at the forefront of the field. It is difficult to situate the included chapters within broad domains, given that

P. L. Hill (✉)
Department of Psychological and Brain Sciences, Washington University in St. Louis, St. Louis, MO, USA
e-mail: patrick.hill@wustl.edu

M. Allemand (✉)
Department of Psychology and University Research Priority Program "Dynamics of Healthy Aging", University of Zurich, Zurich, Switzerland
e-mail: mathias.allemand@uzh.ch

P. L. Hill, M. Allemand (eds.), *Personality and Healthy Aging in Adulthood*, International Perspectives on Aging 26,
https://doi.org/10.1007/978-3-030-32053-9_1

each chapter touches upon both advances in measurement and advances in theory. First, authors will discuss their ongoing efforts to move beyond the personality taxonomies and self-report methods that may have unduly handicapped the precision with which we can predict healthy aging outcomes. Building from this background, researchers from both within and outside psychology present new methods and analytic techniques to add to the researchers' toolbox. Second, across entries, the authors will consider explanatory frameworks that expand upon existing models in order to further our understanding of when, for whom, and why personality constructs predict aging trajectories. Prior to these chapters, though, we first provide a brief overview of the existing knowledge and frameworks on personality and healthy aging and, in so doing, alert the reader to areas of need that will be addressed in the current volume.

1.1 What Is Personality and What Is a Personality Trait?

The definition of personality science comes as a double-edged sword for researchers, as it typically is described as simply "the study of the person" (Funder, 1997). Such a broad definition is advantageous insofar that it allows for a wide array of constructs and individual differences to fall within the umbrella of personality. For instance, one model, known as the neo-socioanalytic framework (Roberts, Wood, & Caspi, 2008; Roberts & Jackson, 2008), outlines that a full account of personality includes assessing an individual's traits, motives, values, cognitive and functional abilities, subjective life narrative, reputation among others, and sense of self and identity. Moreover, this model recognizes that all constructs are inherently contextualized within social roles and cultural expectations, as well as that genetic and physiological mechanisms underlie several of these dimensions. Such an approach to personality science is beneficial as it acknowledges that individual differences across a wide variety of psychosocial variables are important for studying the person. However, it also becomes immediately problematic insofar that no single study can capture all aspects of personality. Thus, it is difficult to ascertain and define the unique contributions of any subset of variables, given that each grouping (traits, motives, abilities, etc.) are inherently intertwined with all other variable clusters.

Accordingly, personality researchers typically focus on one or two subsets of dispositional characteristics, with the knowledge that several aspects of the person must remain unassessed in any given study. Within the realm of health psychology and healthy aging, most researchers have targeted personality traits as the primary dispositional category of interest. Reasons behind this choice have been discussed in greater detail elsewhere, but we focus our discussion here on three primary benefits of trait-based research into healthy aging. First, a wealth of research has focused on demonstrating that personality trait domains often can be found in cultures across the world (John, Soto, & Naumann, 2008), insofar that when we describe the people in our lives, we typically do so by focusing upon similar characteristics or qualities. Second, based on this cross-cultural work, trait taxonomies have been developed that capture the similarities in personality descriptors used worldwide, which have identified five (John et al., 2008) or six (Ashton & Lee, 2007) primary trait dimensions, allowing greater possibilities for research conducted in one study to be replicated or generalized to other contexts. Third, and perhaps most important, personality traits are relatively simple to assess compared to other aspects of personality science, such as life narratives, or constructs that move beyond the individual-level, such as cultural expectations or societally-prescribed roles. Because of these attributes, the trait approach to personality science has been primary for research into health psychology in recent decades (Hampson, 2012), and as such will predominate most chapters in this book.

Another factor contributing to the dominance of the trait approach to personality has been the evolving definition of "trait" in the field. In earlier work, personality traits were viewed as relatively unmalleable constructs whose prediction of behavior varied little across contexts (Costa & McCrae, 1992). More contemporary trait definitions note that personality traits reflect *relatively* stable constructs that hold some influence on behavior across situations and settings (e.g., Roberts, 2009). Indeed, several studies have now demonstrated that normative changes on personality traits occur throughout adulthood (Roberts, Walton, & Viechtbauer, 2006), and individuals hold the capacity for trait change even relatively later in life (Small, Hertzog, Hultsch & Dixon, 2003). Allowing for the possibility of trait change presents immense value for healthy aging researchers because traits are no longer immoveable constructs that may "doom" one to negative outcomes throughout life. Instead, the current zeitgeist is that evidence for personality trait stability and change is available across the lifespan, and the question now is whether we can begin targeting personality trait change in efforts to promote healthy aging outcomes (e.g., Mroczek, 2014).

For all these reasons, there has been a prevalent and potentially undue focus on personality traits as the primary focus of research. A common theme across chapters in the personality measurement section will be identifying the drawbacks to the often overly simplistic efforts to understand the person through the methods common today, such as the typical focus on capturing all individuals' personality profiles using their scores on only five or six trait domains. Though there is inherent value in employing a relatively limited number of domains, chapters in this section will discuss how greater predictive precision for health and aging outcomes will naturally come when targeting "lower-order" or more specific traits. In other words, if the focus is to improve effect size magnitudes and producing aging interventions targeted to specific personality profiles, then we may wish to employ trait taxonomies that allow for greater nuance and specificity at the cost of including longer and more thorough assessment inventories.

1.2 Innovative Methodological and Analytic Techniques for Studying Personality and Healthy Aging

It is difficult to provide an overview for all the measurement and analytic issues present in the field, and thus we focus here on presenting three of the more common critiques of the literature, which will be addressed throughout the chapters in this and other sections of the book. First, and perhaps most common, the field of personality science has been rightly criticized for an undue reliance on self-report methods for understanding the person. Though self-reports are valuable and provide unique information to other methods (Paulhus & Vazire, 2007), there is a clear need to examine alternative methods for capturing personality constructs that are less subject to self-report biases. Similar claims can be made regarding a number of health constructs; self-reported health is a particularly valuable predictor of objective health markers such as later mortality risk (Idler & Benyamini, 1997), but it too can suffer from issues of reporter effects. Accordingly, researchers have sought to address these issues by moving toward alternative approaches to capturing personality and health data, including efforts to understand the person's behavior through frequent measurement assessments. In addition, several chapters will advance these arguments by considering personality dynamics and behavioral manifestations of personality. Indeed, the study of the person needs to account for how every individual interacts with events in daily life, as well as the fact that every individual experiences state-level fluctuations on any given trait across days (e.g., Fleeson, 2001), which will serve as a foundation for discussing current technological advances in the study of personality and healthy aging.

Second, despite the obvious need to move beyond single-sample, single-culture studies, research in the field has been dominated by these types of studies, largely due to the difficulty with assessing participants across multiple countries. This is not to say that single-sample studies have limited value; in fact, they can provide immensely important information on what predicts healthy aging within a given setting. However, before recommending a certain dispositional characteristic, such as being conscientious, as a uniform and widespread promoter of healthy aging, researchers need to compare findings across multiple settings and cultures. This point is made explicit in models of personality discussed above (Roberts et al., 2008) and elsewhere (McAdams & Pals, 2006), and it is particularly important for the study of healthy aging, as the expectations for aging and roles for older adults differ widely across cultures. Accordingly, it would be problematic to suggest that any personality disposition will be uniformly valuable for healthy aging without thorough comparisons across samples from different countries. Though such comparisons are markedly challenging, several chapters herein will discuss analytic approaches toward this end.

Third, a frequent discussion throughout all chapters will be how to actually operationalize "healthy aging." The World Health Organization (WHO) has defined it as "the process of developing and maintaining the functional ability that enables well-being in older age" (World Health Organization, 2019). Though this definition is valuable insofar that it moves beyond the outdated approaches that focus solely on absence of ailments, the WHO definition presents inherent difficulties in capturing all aspects of aging individuals, including their physical mobility, decision-making skills, social contribution, and relationship interactions in daily life. That said, this definition points to the clear relevance of personality science to the study of healthy aging, as personality constructs influence all these domains of life. The question confronting researchers then is, what is "the" outcome of interest when attempting to understand the promotion of healthy aging? Rather than make any recommendations or suggestions on the conceptualization of healthy aging in this introduction, we leave it to the individual chapters to discuss their efforts to address this markedly difficult question.

1.3 Connecting Personality Constructs to Healthy Aging Outcomes

Given that most chapters will focus on traits, we focus our discussion in the remainder of this introduction on personality traits, with the recognition that a primary need for future research is to better address the "other" elements of personality science when investigating links between personality and healthy aging. Personality traits are thought to reflect relatively enduring patterns of thoughts, feelings, and behaviors (Roberts, 2009), and thus researchers are presented with three pathways through which to consider how personality traits predict health and well-being outcomes. Fewer studies have investigated how the cognitions associated with personality dimensions are explanatory mechanisms for personality-health associations (though see Ferguson, 2013, for a theoretical framework that considers cognitive explanatory mechanisms), and more work is needed. Toward this end, multiple chapters in the current volume will focus on how personality traits and personality-related behaviors are associated with cognitive outcomes, with a focus on explaining recent work showing that personality dimensions may prove valuable for predicting normative and non-normative cognitive decline with aging.

Affective pathways linking personality to health have been discussed more frequently in the literature, often focusing on trait neuroticism. Neuroticism can be defined as the tendency to report greater anxiety, depression, and emotional lability in daily life (John et al., 2008). Those who are higher on this trait also experience more negative health outcomes (Hampson, 2012). One potential explanation is that individuals higher on neuroticism experience more negative affect, which then leads to worse health and wellbeing. Similarly, it may be that neuroticism is associated with worse stress reactivity, which in turn leads to problematic aging and health outcomes. These pathways may help explain part of the dramatic economic impact of neuroticism on health care costs at the societal-level (Cuijpers et al., 2010). Affective well-being and its role in personality-aging associations will be discussed across multiple chapters.

Though work has considered affective pathways linking personality to healthy aging, the primary focus in the field thus far has been on behavioral mechanisms. Health psychologists have focused on understanding how to promote healthier lifestyle behaviors largely since the inception of that field, and it is perhaps the most intuitive route by which to understand why personality influences health and well-being. Indeed, any demonstration that a variable predicts who is more likely to smoke or exercise has clear ramifications for understanding why that variable is associated with better or worse health outcomes. Accordingly, the "classic" models linking personality to health often focused on behavioral explanations. For instance, Adler and Matthews (1994) posited that health behavior serves as a central mediator between personality traits and health outcomes, and also recognized that social behaviors and activities (another trait-related behavior) can provide an indirect route to health. More contemporary models also place a central focus on behavioral explanations linking personality to health, outlining a wide variety of important health-relevant actions potentially influenced by one's personality (Ferguson, 2013).

Spurred by these and other models, a wealth of studies have demonstrated that personality traits are consistent predictors of the frequency of health behavior enactment (for a review, see Hampson, 2012). In this discussion, the trait of conscientiousness has taken center stage, a trait that is defined as a propensity toward being self-controlled, organized, and industrious (Roberts, Jackson, Fayard, Edmonds, & Meints, 2009). Meta-analytic work demonstrates that conscientious individuals report a greater likelihood of reducing negative health behaviors, such as smoking and drug use, and are more likely to enact positive health behaviors, such as activity engagement and better diet (Bogg & Roberts, 2004). Accordingly, theoretical and empirical work has presented the case for conscientiousness as a central concern for public health professionals (Bogg & Roberts, 2013).

A central theme for several chapters across this volume will be to build upon these models linking personality to healthy aging, particularly with a focus on how to contextualize them within a lifespan developmental perspective. All too often, work in the field has focused on relatively simplistic explanations, such as linking personality to a single behavioral mechanism as an explanation for effects on health or healthy aging. Moreover, several studies fail to even formally test mediation to understand whether indirect effects through tested mechanisms provide explanatory value for later healthy aging. With respect to making theoretical advances for the literature, the current chapters will take up the aims of (a) linking personality to multiple behavioral, affective, and cognitive explanatory mechanisms, (b) developing the arguments for how these effects play out across the lifespan, and (c) contextualizing these models within cultural or developmental settings.

1.4 Conclusion

In sum, we hope to have provided the case for employing personality science in the study of healthy aging, while introducing some of the primary conversations and difficulties in the field today. In the three sections to follow, we have collected the thoughts from central and up-and-coming researchers in the field, in order to spark new discussions and avenues for future research. Though the questions presented above are complicated and difficult, we hope that the chapters in the current volume motivate the reader to continue addressing these issues, by providing innovative and unique approaches to tackling a central concern in today's increasingly aging society, namely: How can we help individuals maintain their functioning and well-being as they continue adding years to their lives?

References

Adler, N., & Matthews, K. (1994). Health psychology: Why do some people get sick and some stay well? *Annual Review of Psychology, 45*, 229–259.

Ashton, M. C., & Lee, K. (2007). Empirical, theoretical, and practical advantages of the HEXACO model of personality structure. *Personality and Social Psychology Review, 11*, 150–166.

Bogg, T., & Roberts, B. W. (2004). Conscientiousness and health-related behaviors: A meta-analysis of the leading behavioral contributors to mortality. *Psychological Bulletin, 130*, 887–919.

Bogg, T., & Roberts, B. W. (2013). The case for conscientiousness: Evidence and implications for a personality trait marker of health and longevity. *Annals of Behavioral Medicine, 45*(3), 278–288.

Costa, P. T., Jr., & McCrae, R. R. (1992). Four ways five factors are basic. *Personality and Individual Differences, 13*, 653–665.

Cuijpers, P., Smit, F., Penninx, B. W., de Graaf, R., ten Have, M., & Beekman, A. T. (2010). Economic costs of neuroticism: A population-based study. *Archives of General Psychiatry, 67*(10), 1086–1093.

Ferguson, E. (2013). Personality is of central concern to understand health: Towards a theoretical model for health psychology. *Health Psychology Review, 7*, S32–S70.

Fleeson, W. (2001). Toward a structure-and process-integrated view of personality: Traits as density distributions of states. *Journal of Personality and Social Psychology, 80*, 1011–1027.

Funder, D. C. (1997). *The personality puzzle*. New York, NY: W. W. Norton & Co..

Hampson, S. E. (2012). Personality processes: Mechanisms by which personality traits "get outside the skin". *Annual Review of Psychology, 63*, 315–339.

Idler, E. L., & Benyamini, Y. (1997). Self-rated health and mortality: A review of twenty-seven community studies. *Journal of Health and Social Behavior, 38*, 21–37.

John, O. P., Naumann, L. P., & Soto, C. J. (2008). Paradigm shift to the integrative big five trait taxonomy. In O. P. John, R. W. Robins, & L. A. Pervin (Eds.), *Handbook of personality: Theory and research* (3rd ed., pp. 114–158). New York, NY: Guilford Press.

McAdams, D. P., & Pals, J. L. (2006). A new big five: Fundamental principles for an integrative science of personality. *American Psychologist, 61*, 204–217.

Mroczek, D. K. (2014). Personality plasticity, healthy aging, and interventions. *Developmental Psychology, 50*, 1470–1474.

Paulhus, D. L., & Vazire, S. (2007). The self-report method. In R. W. Robins, R. C. Fraley, & R. F. Krueger (Eds.), *Handbook of research methods in personality psychology* (pp. 224–239). New York, NY: Guilford Press.

Roberts, B. W. (2009). Back to the future: Personality and assessment and personality development. *Journal of Research in Personality, 43*, 137–145.

Roberts, B. W., & Jackson, J. J. (2008). Sociogenomic personality psychology. *Journal of Personality, 76*(6), 1523–1544.

Roberts, B. W., Jackson, J. J., Fayard, J. V., Edmonds, G., & Meints, J. (2009). Conscientiousness. In M. R. Leary & R. H. Hoyle (Eds.), *Handbook of individual differences in social behavior* (pp. 369–381). New York, NY: The Guilford Press.

Roberts, B. W., Walton, K. E., & Viechtbauer, W. (2006). Patterns of mean-level change in personality traits across the life course: A meta-analysis of longitudinal studies. *Psychological Bulletin, 132*, 1–25.

Roberts, B. W., Wood, D., & Caspi, A. (2008). The development of personality traits in adulthood. In O. P. John, R. W. Robins, & L. A. Pervin (Eds.), *Handbook of personality: Theory and research* (3rd ed., pp. 375–398). New York, NY: Guilford Press.

Siegman, A. W. (1994). Cardiovascular consequences of expressing and repressing anger. In A. W. Siegman & T. W. Smith (Eds.), *Anger, hostility, and the heart* (pp. 173–197). Hillsdale, NJ: Erlbaum.

Small, B. J., Hertzog, C., Hultsch, D. F., & Dixon, R. A. (2003). Stability and change in adult personality over 6 years: Findings from the Victoria longitudinal study. *The Journals of Gerontology Series B: Psychological Sciences and Social Sciences, 58*(3), P166–P176.

World Health Organization. (2019). *What is healthy ageing?* Retrieved from https://www.who.int/ageing/healthy-ageing/en/

Chapter 2
Integrating Personality and Relationship Science to Explain Physical and Mental Health

Hannah Brazeau and William J. Chopik

In traditional vows, married couples often make the promise to care for one another "in sickness and in health". This vow expresses that romantic partners should remain committed to each other regardless of the obstacles that life puts in their way, including when one member of the relationship has compromised health. However, this vow seems to suggest that ill health is a condition that develops and occurs outside the context of a romantic relationship, which a couple must then manage as a unit. In the current chapter, we will highlight how this could not be further from the truth. In fact, an individual's mental and physical health can depend on the quality of these social relationships. But how exactly do these processes occur? We will argue that the personality characteristics that each partner brings to a relationship play a role in shaping how an individual interprets and experiences their relationships, which inevitably influences one's health. Although there are large literatures examining the associations of health with personality and interpersonal relationships independently, there are also many opportunities for these two areas of psychology to intersect in an attempt to explain the health consequences of romantic relationships as they occur across the lifespan.

In the current chapter, we describe how personality and close relationship processes may interact to influence mental and physical health. We begin with a discussion of how our romantic relationships contribute to our health and how personality can predict some of the relationship outcomes that are important in this connection. Next, we showcase some of the prominent models enabling researchers to characterize how personality and relationship factors may interact to influence health. We close with a discussion of the unanswered questions that will help to direct future

H. Brazeau (✉)
Department of Psychology, Carleton University, Ottawa, ON, Canada
e-mail: HannahBrazeau@cmail.carleton.ca

W. J. Chopik
Department of Psychology, Michigan State University, East Lansing, MI, USA

P. L. Hill, M. Allemand (eds.), *Personality and Healthy Aging in Adulthood*, International Perspectives on Aging 26,
https://doi.org/10.1007/978-3-030-32053-9_2

research examining the combined impact that personality and relationships has on health.

2.1 How Do Our Romantic Relationships Impact Our Mental and Physical Health?

Before discussing how it is that personality and romantic relationships may interact to influence health, we must first demonstrate that: (1) our romantic relationships play an important role in determining physical and mental health, and (2) personality plays a role in determining the behaviours and experiences people have in their romantic relationships that are important to the relationship-health link. In this section, we will address the first point by describing the impact that our romantic relationships have on physical and mental health through-out the lifespan before outlining how it is that these relationships have this effect.

For decades, researchers have argued that social relationships and interactions are a basic human need that is crucial to living a happy and healthy life (e.g., Baumeister & Leary, 1995; Holt-Lunstad, Smith, & Layton, 2010). Romantic relationships are often used to vouch for this argument as those involved in a committed romantic relationship generally live longer, healthier and more satisfying lives than their noncommitted peers (Bennett, 2006; Dupre, Beck, & Meadows, 2009; Rogers, 1995). In particular, individuals in romantic relationships tend to report considerably better self-reported physical health (Lui & Umberson, 2008; Rohrer, Bernard, Zhang, Rasmussen, & Woroncow, 2008; Umberson, 1992), as well as better emotional well-being and greater life satisfaction (Bookwala & Schultz, 1996; Gove & Tudor, 1973; Horwitz, White, & Howell-White, 1996; Kessler & Essex, 1982; Tucker, Friedman, Wingard, & Schwartz, 1996; Wadworth, 2016). These effects are especially large when comparing married individuals to those who are widowed and divorced (compared to single), as the breaking of relationship bonds can have strong negative impacts on self-reported physical and mental health (Rook & Zettel, 2005; Williams & Umberson, 2004). Older adulthood is a period of the lifespan in which this association is especially critical as widowhood is typical in this age-group and older adults generally tend to already have poorer health when compared to younger adults. However, perhaps one of the most significant health benefits associated with being in a committed relationship is the minimized probability of developing a variety of acute and chronic physical and psychological conditions (Datta, Neville, Kawachi, Datta, & Earle, 2009; Nilsson, Engstrom, & Hedblad, 2008; Umberson, Williams, Powers, Liu, & Needham, 2006). This includes a substantially lower morbidity and mortality risk for cardiovascular disease and cancer, which represent two of the leading causes of death in North America (Centers for Disease Control and Prevention, 2017; Canada, 2015), as well as lower risk of anxiety and mood disorder diagnosis (see Umberson & Williams, 1999; Waite & Gallagher, 2000, for reviews), which are among the most common mental health disorders. In sum, there is a substantial body of research indicating that being involved in a romantic relationship can be beneficial for both physical and mental health. But how is it that romantic relationships have these effects on our health?

Of course, it is not merely an individual's relationship status that impacts health, instead it is the experiences within and the quality of these relationships that influence health status (Gottman & Notarius, 2002). Indeed, many theoretical models linking relationships and health propose that the behaviours and outcomes experienced within a relationship are essential components in predicting health outcomes (Kiecolt-Glaser & Newton, 2001; Pietromonaco, Uchino, & Dunkel Schetter, 2013). This notion is supported by research indicating that having a happier and more satisfying relationship tends to coincide with living a happier and healthier life in all age groups. Specifically, those in satisfying relationships tend to report having better physical health and fewer health ailments (Bookwala, 2005; Miller, Dopp, Myers, Stevens, & Fahey, 1999; Robles, Slatcher, Trombello, & McGinn, 2014; Wickrama, Lorenz, Conger, & Elder, 1997), as well as greater psychological well-being and fewer depressive symptoms than those who are relatively unsatisfied in their relationships (Proulx, Helms, & Buehler, 2007; Whisman, 2001). Beyond relationship satisfaction, positive relationship experiences (e.g., social support, intimacy, physical touch) also have beneficial effects on physical and mental health. These positive experiences are said to alleviate the effect of stress on various psychosocial and physiological pathways that influence health (e.g., Slatcher & Selcuk, 2017). For instance, romantic partners experience lower cortisol levels on days when they engage in more physical touch (i.e., holding hands, hugging) with their spouses (Ditzen, Hoppmann, & Klumb, 2008). However, not all relationships are classified as being satisfying or characterized by positive relationships experiences. So the question becomes, when an individual is involved in an unsatisfying relationship, what happens to their physical and mental health?

As you may have expected, just as a happy and well-adjusted relationship is beneficial to health, an unhappy or poorly functioning romantic relationship can be harmful to health (Robles & Kiecolt-Glaser, 2003). In fact, individuals who are not satisfied in their romantic relationships are more likely to report experiencing a variety of physical and mental health conditions including cardiovascular disease, anxiety disorders, and depression (Frech & Williams, 2007; Hawkins & Booth, 2005; Overbeek et al., 2006). Further, negative relationship experiences (e.g., anger, relationship conflict, hostility, criticism) have also been shown to undermine health (Bookwala, 2005; Choi & Marks, 2008). This may occur because problematic social interactions can evoke negative psychological and physiological responses, which if chronically activated are associated with future health difficulties. For instance, negative relationship experiences, such as conflict and relationship strain, are associated with physiological markers of stress and detriments in immune system functioning that undermine later physical health (Kiecolt-Glaser, 2018; Kiecolt-Glaser et al., 2005; Miller et al., 1999; Robles & Kiecolt-Glaser, 2003). Similarly, these negative relationship interactions are associated with psychological distress and depression, which can have adverse impacts on long-term mental health (Fincham & Beach, 1999; Proulx et al., 2007).

The breadth of research reviewed above communicates the substantial impact that our romantic relationships have on our physical and mental health. In particular, we have outlined how relationship quality and the experiences that one has within a romantic relationship influences whether the relationship will be a benefit or a detriment to one's health. But what are the factors that determine whether an individual will have a satisfying and functional romantic relationship? To answer this question, we will now examine the ways in which personality affects relationships and relationship quality.

2.2 Can Personality Determine Who Flourishes or Flounders Within a Relationship?

The previous section demonstrated that the behaviours and experiences that one has within a romantic relationship have a substantial impact on one's health. However, that was only one piece of the puzzle as we also specified that we had to demonstrate that personality can determine the behaviours and experiences that an individual will likely have in their romantic relationships. In this section, we will discuss how two theories of personality can be used to influence the relationship behaviours and outcomes that we just demonstrated have considerable impact on physical and mental health.

Often our personalities play a role in how we interpret and behave within interpersonal situations. Thus, not surprisingly, personality traits are suggested to predict relationship quality, relationship experiences, relationship dissolution, and marital divorce (Roberts, Kuncel, Shiner, Caspi, & Goldberg, 2007; Robins, Caspi, & Moffitt, 2002). In fact, it is estimated that up to 60% of the variance in marital quality and 25% of the variance in divorce risk can be explained by the personality traits of the spouses involved in the relationship (Jocklin, McGue, & Lykken, 1996; Russell & Wells, 1994; Solomon & Jackson, 2014). The research linking personality to relationship experiences has primarily focused on the impact that constructs from *attachment theory* and *the Big Five model* have on relationship behaviours and outcomes. Although we acknowledge the large literature investigating other individual differences in relationship research (e.g., self-esteem, narcissism; Murray, Rose, Bellavia, Holmes, & Kusche, 2002; Brunell & Campbell, 2011), we will concentrate on discussing the influence that attachment and the Big Five personality traits have on the relationship experiences that we previously established were associated with health. However, before we begin, it should be noted that since these individual difference factors are believed to be relatively stable over time, the impacts that attachment and the Big Five traits have on relationship processes tends to be fairly stable across relationships and the lifespan.

2.2.1 Adult Attachment

Attachment theory is one of the only prominent theories of personality that was designed with interpersonal interactions specially in mind. The original purpose of attachment theory was to describe and explain the close, emotional bond that develops between an infant and his or her primary caregiver (Bowlby, 1969). However, it was quickly expanded to describe adulthood relationships as the attachment processes responsible for the bonds that develop between adults were deemed to be similar to the ones responsible for the bond that develops between an infant and caregiver (Bowlby, 1969; Fraley & Shaver, 2000; Hazan & Shaver, 1994). Regardless as to whether we are referring to children or adults, the underlying notion behind attachment theory is the same: individuals develop an *attachment orientation*—patterns of interpersonal cognitions, emotions, and behaviors—based on their unique interactions and experiences with attachment figures. It is these attachment orientations that guide how an individual interprets and behaves within their close relationships (Fraley & Shaver, 2000; Shaver & Mikulincer, 2007). In adulthood, people are thought to vary on two independent dimensions of attachment, which determine their attachment orientation: (a) attachment anxiety, which refers to the tendency to ruminate and be obsessively worried about close relationships due to fears of rejection and abandonment, and (b) attachment avoidance, which involves the tendency to experience discomfort in situations of physical and emotional closeness or dependence (Brennan, Clark, & Shaver, 1998; Campbell & Marshall, 2011; Fraley & Shaver, 2000). Individuals who report high levels of either attachment anxiety or avoidance are said to display an insecure attachment orientation, whereas those who report low levels on both dimensions are thought to exhibit attachment security, which refers to the tendency to feel comfortable with interpersonal closeness as well as independence. Now that the basis of attachment theory has been established, we can discuss how each attachment orientation can shape the relationship experiences that are significant to the connection between relationships and health. In particular, we will focus on relationship quality (i.e., relationship satisfaction and commitment) and stability.

When evaluating the impact that attachment has on relationships, researchers often focus on whether or not people are happy with and committed to their partner (i.e., relationship quality; Etcheverry, Le, Wu, & Wei, 2013). This focus has consistently demonstrated that individuals higher on attachment insecurity experience lower levels of relationship satisfaction in romantic relationships compared to those with greater attachment security (see Mikulincer & Shaver, 2016 for detailed review). In fact, these individuals report lower daily relationship satisfaction (Campbell, Simpson, Boldry, & Kashy, 2005; Lavy, Mikulincer, & Shaver, 2013; Neff & Karney, 2009), and tend to be less satisfied with their relationships in the first 3 years of marriage (Davila, Karney, & Bradbury, 1999). The negative impact that attachment insecurity has on romantic relationships is not limited to relationship satisfaction as similar patterns have been seen for relational commitment. Specifically, those who exhibit attachment insecurity often report being less committed to their romantic partner (see Mikulincer & Shaver, 2016 for detailed review). Given that relationship quality is one of the largest predictors of relationship stability, individuals with attachment insecurity also tend to be in relationships that are characterized by instability. In fact, anxious and avoidant individuals tend to have shorter dating relationships and marriages when compared to more secure individuals (Birnbaum, Orr, Mikulincer, & Florian, 1997; Crowell & Treboux, 2001; Duemmler & Kobak, 2001; Kirkpatrick & Davis, 1994). However, the reason for this instability differs according to the individual's attachment orientation. In particular, despite being relatively unsatisfied and uncommitted to their relationships, anxious individuals are overly dependent on the affections of their partner, while also having a generalized fear of abandonment. Thus, these individuals are at greater risk of remaining in unhappy marriages for extended periods of time (Davila & Bradbury, 2001; Kirkpatrick & Davis, 1994). But these relationships are often unstable as anxious individuals commonly breakup and re-partner with the same people multiple times. On the other hand, since avoidant individuals have an overall dislike of emotional intimacy, these individuals have a higher likelihood of terminating a relationship as soon as relationship distress is experienced (Kirkpatrick & Davis, 1994).

The impact that attachment insecurity has on relationship quality and stability may be understood by the behavioural tendencies associated with each of the insecure attachment orientations. For instance, due to their fears of rejection and abandonment, anxious individuals require constant affection and reassurance from their partner in order to feel a sense of emotional closeness and stability (Hazan & Shaver, 1994; Shaver & Mikulincer, 2007). This results in these individuals consistently engaging in behavioral and emotional strategies that involve seeking proximity and closeness to a romantic partner (i.e., hyperactivating strategies; Mikulincer & Shaver, 2012). However, they are often disappointed by their partners' lack of reciprocity of affection, which has been suggested to lead to low levels of satisfaction and commitment among anxious individuals. In comparison, because avoidant individuals are uncomfortable with closeness or dependence, these individuals often attempt to maintain a sense of autonomy while in a romantic relationship (Hazan & Shaver, 1994; Shaver & Mikulincer, 2007). Thus, avoidant individuals engage in behavioral and emotional strategies that distance themselves from a romantic partner (i.e., deactivating strategies). These include being more critical of their partner, tending not turn to their partner for support, and ignoring the emotional cues and signals in their relationship (Campbell et al., 2005; Simpson, Rholes, & Nelligan, 1992). It is these dismissive tendencies that are believed to account for the consistently low levels of relationship satisfaction and commitment reported by individuals high in avoidance.

Although attachment theory takes a relationship perspective on individual differences and applies it to social relationships, some researchers still prefer to take a broader personality approach when examining the impact that personality has on our relationships. Thus, many researchers have examined the impact that the Big Five traits have on relationship experiences and outcomes. Despite having some empirical overlap with attachment orientations (e.g., Noftle & Shaver, 2006), an entirely separate literature examining the associations between the personality traits outlined in the Big Five model and relationship functioning has emerged.

2.2.2 The Big Five Model

The Big Five model is perhaps the most popular and well-known theory of personality. The five personality traits that make up this model—agreeableness, conscientiousness, neuroticism, extraversion, and openness—have emerged as an empirically based framework that captures the major individual differences that exist between people (John & Srivastava, 1999). Although not developed with interpersonal relationships specifically in mind, the personality traits outlined in the Big Five nevertheless impact how we understand, experience and act within our social relationships. Below, we will describe each of the Big Five personality traits and outline how these traits can shape relationship quality and stability, which we have been shown to be related to health.

Agreeableness is a social trait that reflects individual differences in the propensity to be altruistic, trusting, modest, and warm (John & Srivastava, 1999). Agreeable individuals are more motivated to maintain positive relationships with others and they tend to engage more in social behaviors that facilitate intimacy (Jensen-Campbell & Graziano, 2001). Not surprisingly, agreeableness has been consistently related to many positive relationship variables. For instance, agreeableness is positively associated with marital and relationship satisfaction, and negatively associated with both marital dissatisfaction and divorce (Dyrenforth, Kashy, Donnellan, & Lucas, 2010). In addition, agreeable individuals experience greater intimacy, commitment, and passion within their relationships (Ahmetoglu, Swami, & Chamorro-Premuzic, 2010). Although agreeableness is linked to relationship quality, little research has directly examined *why* the two are linked. Some researchers have suggested that the link may be explained by the impact that agreeableness has on conflict. Generally speaking, agreeable individuals tend to report having fewer negative interactions with their romantic partners (Donnellan, Conger, & Bryant, 2004), and use more productive conflict resolution skills when negative interactions present themselves in a relationship (Graziano, Jensen-Campbell, & Hair, 1996).

Conscientiousness is a trait of self-discipline that reflects individual differences in the propensity to plan, organize, delay gratification, and be achievement-oriented (Roberts, Jackson, Fayard, Edmonds, & Meints, 2009a). Conscientiousness has been found to be positively related to relationship satisfaction in both dating and married couples (Dyrenforth et al., 2010; Engel, Olson, & Patrick, 2002). Further, it is positively associated with many other positive relationship experiences such as commitment, passion, and intimacy (Ahmetoglu et al., 2010; Engel et al., 2002). It is thought that the achievement orientation of those with a conscientious personality motivate these individuals to make their relationships work; while their other tendencies allow them to consistently enact positive relationship behaviors, such as remembering relevant information, being reliable, and upholding promises (Jackson et al., 2010). Further, the self-control habits of conscientious individuals are believed to prevent relationship problems as these individuals can easily avoid temptations, such as enacting revenge and committing infidelity (Engel et al., 2002). The responsible, dependable, and hardworking nature of those with a conscientious personality generally creates fewer areas of disagreement (Donnellan et al., 2004). However, future work is needed to explicitly test the mechanisms that link conscientiousness to these positive relationship experiences and outcomes.

Neuroticism reflects individual differences in the tendency to experience negative emotions and emotional instability (John & Srivastava, 1999). Early personality researchers identified that many of these tendencies and emotional experiences create and define unhappy and unstable relationships (Karney & Bradbury, 1995). A significant body of research has supported this insight by linking neuroticism with poor relationship quality (e.g., Barelds, 2005; Davila, Karney, Hall, & Bradbury, 2003; Donnellan et al., 2004), and a greater risk of marital dissolution (Kurdek, 1993; Roberts et al., 2007). In fact, neuroticism has been found to have the strongest association with marital satisfaction when compared to the other four Big Five personality traits (Heller, Watson, & Ilies, 2004). This is also found when examining the effects of partner personality—having a highly neurotic partner is among the largest predictors of life and relationship satisfaction (Chopik & Lucas, 2019). These results may be explained by the tendencies of neurotic individuals to express more criticism, contempt, and defensiveness, which may damage a relationship. Further, these individuals use greater avoidance coping (Lee-Baggley, Preece, & DeLongis, 2005) and report experiencing fewer positive and more negative social interactions overall (Russell & Wells, 1994). Such characteristics may lead to conflicts in a romantic relationship and prevent others from engaging in socially supportive behaviors. Overall, when comparing the Big Five personality traits, neuroticism is generally the strongest and most consistent trait impacting our social relationships. However, unlike agreeableness and conscientiousness, neuroticism is characterized by lower relationship quality and more negative relationship experiences.

Like agreeableness, extraversion has a strong sociability component and refers to individual differences in the tendencies to be sociable, active, assertive, and to experience positive affect (John & Srivastava, 1999). It is often believed that extraverted individuals should have more positive relationship experiences because these individuals are better equipped to communicate their desires, wants, and intentions than those low on extraversion (Taraban, Hendrick, & Hendrick, 1998). However, the available literature on the associations between extraversion and relationship outcomes has yielded inconsistent findings (e.g., Donnellan et al., 2004; Watson, Hubbard, & Wiese, 2000). For instance, extraversion has been associated with several positive relationship variables, such as satisfaction (Solomon & Jackson, 2014) and passion (Ahmetoglu et al., 2010). But it has also been associated with many negative relationship variables, such as lower relationship satisfaction (for men; Bentler & Newcomb, 1978) and marital instability (Ahmetoglu et al., 2010). The inconsistent findings regarding extraversion may be explained by the fact that extraversion appears to capture individual differences that relate to social impact (as opposed to maintaining positive relations), which may not be an important process in shaping romantic relationships (Tobin, Graziano, Vanman, & Tassinary, 2000). Regardless, given the empirical inconsistencies concerning the role of extraversion in close relationships, future work will be needed to create a clearer picture.

Openness reflects individual differences in the propensity to be imaginative, creative, curious, and adventurous (John & Srivastava, 1999). Originally, it was thought that openness may facilitate relationship quality and maintenance by promoting intellectual approaches to problem solving, more flexible attitudes towards change, and a willingness to try new things within a relationship (Robins et al., 2002). However, it appears that openness is often unassociated or has conflicting associations with relationship quality (e.g., Chopik & Lucas, 2019). Meta-analyses examining the impact of personality on relationship quality has demonstrated that there is often no relation between openness and relationship satisfaction (Malouff, Thorsteinsson, Schutte, Bhullar, & Rooke, 2010). Further, openness has not been found to have a significant association with other relationship factors, including intimacy, passion or commitment (Engel et al., 2002). Thus, openness may not be the most important individual difference characteristics to consider when attempting to understand relationship processes.

The research reviewed above implies that individual differences in attachment and the Big Five traits help to explain variability in how individuals interpret, behave, and experience their romantic relationships. We will now combine this discussion with that outlining the ways in which relationship processes influence health so that we may consider how it is that personality and relationship experiences may interact to influence physical and mental health.

2.3 How Do Individual Differences in Personality Interact with Relationship Processes to Influence Health?

The first section of the current chapter focused on establishing the role that various relationship processes have on physical and mental health. We followed this with a discussion of how personality plays a role in determining the relationship processes that are important to the relationship-health link. However, the independent discussion of these two topics defeats the overall purpose of the current chapter to consider how it is that personality and relationship science may work together to impact health across the lifespan. Thus, in this section, we will work towards outlining how it is that the research reviewed in the two previous sections may be combined to provide us with a better understanding of how our relationships and personality traits impact our mental and physical health.

The notion of combining relationship and personality science in an effort to better understand health is not a completely novel idea. For instance, Pietromonaco et al. (2013) developed a model outlining how individual differences in attachment interact with relationship processes to influence physical health. The model suggests that differences in attachment help determine how an individual behaves in a relationship, which subsequently impacts how that relationship functions. These relationship behaviours and outcomes shape physiological stress response patterns, affect, and health behavior engagement, which contribute to later physical health and disease outcomes. Similarly, Slatcher and Selcuk (2017) designed a broader model to explain how personality and marital experiences interact to impact physical health. Within this model, individual difference characteristics (not limited to attachment) influence the positive (i.e., strengths) and negative (i.e., strain) aspects of a relationship. These aspects are thought to impact physical health by influencing various psychological (i.e., affect and cognitions) and behavioural (i.e., health behaviours) mechanisms. Although separate models, these two frameworks tell the same story: personality plays a role in determining various relationships processes, which subsequently influence health and disease outcomes.

Unfortunately, there is very limited evidence to support theoretical frameworks like those designed by Pietromonaco et al. (2013), and Slatcher and Selcuk (2017). In fact, the state of the current literature follows the same structure as the current chapter: there are the researchers who focus on the associations between relationships and health, and then there are a variety of other researchers linking personality to relationship processes and health. And, it appears as though many of the personality and relationship researchers working in the area of health psychology ignore the existence of one another. The emphasis that is placed on examining either the influence that personality or relationship processes have on health creates an incomplete understanding of various important health outcomes. In other words, although comprehensive models have been developed, we are still unsure as to how personality characteristics and relationship processes may be working together to influence health. Despite the lack of research combining these two areas of psychology, there are two ways in which personality and relationship processes may be combined to influence health. The first is that individual difference characteristics *moderate* the link between relationship processes and health (as seen in Slatcher & Selcuk, 2017). That is to say that the extent to which our relationships impact our health is dependent on our personality. This pathway has found some support as relationship conflict has been shown to be related to disruptions in immune systems responses, particularly among highly anxious individuals (Powers, Pietromonaco, Gunlicks, & Sayer, 2006). Broadly, this research suggests that the impact that negative relationship interactions have on health is more prominent for individuals with certain personality traits, in this case individuals who are anxiously attached. But overall, there is a lack of studies examining the possible moderators for the links between relationships and health (Robles et al., 2014).

The second way in which personality and relationship processes may be combined to influence health is that relationship experiences may *mediate* the link between personality and health (as seen in Pietromonaco et al., 2013). This pathway takes a slightly different approach as it suggests relationship processes are able to explain the association between personality and health. In other words, *mediation* implies that personality does not have a direct effect on health. Instead personality influences health via the relationship processes that personality impacts. As we have demonstrated in the two previous sections of this chapter, there is piecemeal support for the mediation argument. That is, personality plays a role in determine various relationship processes, which later have an impact on physical and mental health. However, this is insufficient support and there are few studies that have examined both personality and relationships experiences together in an effort to establish whether relationship processes can explain the association between individual difference characteristics and health.

In sum, a handful of researchers have clearly outlined how it is that relationship processes and personality may interact to influence physical and mental health. We reviewed two models which each take a different approach in combining relationship and personality science to explain health. Although theoretically sound, these models are lacking in empirical support. But in order to test these models correctly, researchers need to ensure that the data and the analytics strategy are sufficient for such an undertaking. Thus, we will now discuss the ways in which we think researchers should be testing these models and answering the questions posed in the current chapter.

2.4 Future Directions

When conducting research including romantic partners it is important to remember that relationships are dynamic and reciprocal. That is, the reactions, experiences, and traits of one partner influence and are influenced by those of the other partner. This mutual influence between romantic partners becomes apparent when examining models such as the one designed by Pietromonaco et al. (2013) as the impact that each partner has on one another is clearly delineated within their framework. Thus, in order to formally test the mechanisms through which individual differences and close relationships influence health, researchers need to design studies in a way that allows for an assessment of both partners' characteristics and outcomes (i.e., dyadic research). In particular, the data analytics that are needed for dyadic studies require an adjustment for non-independence as partners' responses are often impacted by one another. Further, the study methods and analyses need to account for how each person's own characteristics, those of their partner, and the interaction between their own and their partner's characteristics predict the various outcomes of interest (see Kenny, Kashy, & Cook, 2006). The Actor–Partner Interdependence Model (APIM; see Kenny et al., 2006; Kenny, 2018) is a popular analytic model that enables researchers to test the each of these pathways. In particular, APIM allows researchers to examine the extent to which: (a) the characteristics of each relationship partner influence their own outcomes (i.e., actor effects), (b) the characteristics of one relationship partner influence the other partner's outcomes (i.e., partner effects), and (c) the characteristics of one relationship partner interact with those of the other partner to predict one or both partners' outcomes (i.e., interactive actor x partner effects). To date, many dyadic studies that include individual difference measures have tried to merely establish links between the characteristics of each spouse to each spouse's health status (Kim, Chopik, & Smith, 2014; Roberts, Smith, Jackson, & Edmonds, 2009b). These studies ignore the possible relationship processes that may act as mechanisms for the associations between personality and health, such as those outlined in the theoretical frameworks designed by Pietromonaco et al. (2013) and Slatcher and Selcuk (2017). Thus, much more dyadic research is needed to establish how it is that personality characteristics and relationships experiences of each partner may work together to impact the health status of each of the partners in a romantic relationship. Beyond conducting dyadic research, it is necessary for health research to be conducted longitudinally as many health outcomes develop over time after many social interactions with a romantic partner. Naturally, this adds to the complexity of the models discussed above, as now the dyadic analytics also need to model how personality and relationship processes unfold over time to influence the health conditions that develop across the lifespan (Ledermann & Kenny, 2017).

In sum, it is important for future research to ensure that: (a) dyadic longitudinal data is collected on personality, relationship processes, and health outcomes, and (b) the analytic strategy used allows for researchers to correctly address all the different factors that play a role in deciding how the relationship processes and personality of each partner may work together to influence physical and mental health.

2.5 Conclusion

At the beginning of this chapter, we argued that the personality characteristics each partner brings to a relationship play a role in shaping how an individual interprets and experiences their relationships, which inevitably influences one's health. We believe that the current chapter has outlined a solid theoretical basis for this argument. However, the current piecemeal empirical evidence that exists to support this argument is not sufficient to glean strong conclusions regarding the value of the discussed theories. Thus, it is our hope that this chapter provides future researchers with the necessary theoretical and practical knowledge to test this argument more thoroughly as we believe that integrating the science of individual differences with relationship science can bring us closer to the prospect of living happily—and healthily—ever after.

References

Ahmetoglu, G., Swami, V., & Chamorro-Premuzic, T. (2010). The relationship between dimensions of love, personality, and relationship length. *Archives of Sexual Behavior, 39*(5), 1181–1190.

Badr, H., Laurenceau, J. P., Schart, L., Basen-Engquist, K., & Turk, D. (2010). The daily impact of pain from metastatic breast cancer on spousal relationships: A dyadic electronic diary study. *Pain, 151*(3), 644–654.

Barelds, D. P. H. (2005). Self and partner personality in intimate relationships. *European Journal of Personality, 19*(6), 501–518.

Baumeister, R. F., & Leary, M. R. (1995). The need to belong: Desire for interpersonal attachments as a fundamental human motivation. *Psychological Bulletin, 117*(3), 497–529.

Bennett, K. M. (2006). Does marital status and marital status change predict physical health in older adults? *Psychological Medicine, 36*, 1313–1320.

Bentler, P. M., & Newcomb, M. D. (1978). Longitudinal study of marital success and failure. *Journal of Consulting and Clinical Psychology, 46*(5), 1053–1070.

Birnbaum, G. E., Orr, I., Mikulincer, M., & Florian, V. (1997). When marriage breaks up: Does attachment style contribute to coping and mental health? *Journal of Social and Personal Relationships, 14*(5), 643–654.

Bogg, T., & Roberts, B. W. (2004). Conscientiousness and health-related behaviors: A meta-analysis of the leading behavioral contributors to mortality. *Psychological Bulletin, 130*(6), 887–919.

Bookwala, J. (2005). The role of marital quality in physical health during the mature years. *Journal of Aging and Health, 17*(1), 85–104.

Bookwala, J., & Schultz, R. (1996). Spousal similarity in subjective well-being: The cardiovascular health study. *Psychology and Aging, 11*(4), 586–590.

Bowlby, J. (1969). *Attachment and loss: Vol. 1: Attachment*. New York, NY: Basic.

Brennan, K. A., Clark, C. L., & Shaver, P. R. (1998). Self-report measurement of adult romantic attachment: An integrative overview. In J. A. Simpson & W. S. Rholes (Eds.), *Attachment theory and close relationships* (pp. 46–76). New York, NY: Guilford Press.

Brunell, A. B., & Campbell, W. K. (2011). Narcissism and romantic relationships: Understanding the paradox. In W. K. Campbell & J. D. Miller (Eds.), *The handbook of narcissism and narcissistic personality disorder: Theoretical approaches, empirical findings, and treatments* (pp. 344–350). Hoboken, NJ: Wiley.

Campbell, L., & Marshall, T. (2011). Anxious attachment and relationship processes: An interactionist perspective. *Journal of Personality, 79*(6), 1219–1250.

Campbell, L., Simpson, J. A., Boldry, J., & Kashy, D. A. (2005). Perceptions of conflict and support in romantic relationships: The role of attachment anxiety. *Journal of Personality and Social Psychology, 88*(3), 510–531.

Centers for Disease Control and Prevention. (2017). *Leading causes of death*. Retrieved from https://www.cdc.gov/nchs/fastats/leading-causes-of-death.htm

Choi, H., & Marks, N. F. (2008). Marital conflict, depressive symptoms, and functional impairment. *Journal of Marriage and Family, 70*(2), 377–390.

Chopik, W. J., & Lucas, R. E. (2019). Actor, partner, and similarity effects of personality on global and experienced well-being. *Journal of Research in Personality, 78*, 249–261.

Crowell, J., & Treboux, D. (2001). Attachment security in adult partnerships. In C. Clulow (Ed.), *Adult attachment and couple psychotherapy: The 'secure base' in practice and research* (pp. 28–42). New York, NY: Brunner-Routledge.

Datta, G. D., Neville, B. A., Kawachi, I., Datta, N. S., & Earle, C. C. (2009). Marital status and survival following bladder cancer. *Journal of Epidemiology and Community Health, 63*(10), 807–813.

Davila, J., & Bradbury, T. N. (2001). Attachment insecurity and the distinction between unhappy spouses who do and do not divorce. *Journal of Family Psychology, 15*(3), 371–393.

Davila, J., Karney, B. R., & Bradbury, T. N. (1999). Attachment change processes in the early years of marriage. *Journal of Personality and Social Psychology, 76*(5), 783–802.

Davila, J., Karney, B. R., Hall, T. W., & Bradbury, T. N. (2003). Depressive symptoms and marital satisfaction: Within-subject associations and the moderating effects of gender and neuroticism. *Journal of Family Psychology, 17*(4), 557–570.

Ditzen, B., Hoppmann, C., & Klumb, P. (2008). Positive couple interactions and daily cortisol: On the stress-protecting role of intimacy. *Psychosomatic Medicine, 70*(8), 883–889.

Donnellan, M. B., Conger, R. D., & Bryant, C. M. (2004). The big five and enduring marriages. *Journal of Research in Personality, 38*(5), 481–504.

Duemmler, S. L., & Kobak, R. (2001). The development of commitment and attachment in dating relationships: Attachment security as relationship construct. *Journal of Adolescence, 24*(3), 401–415.

Dupre, M. E., Beck, A. N., & Meadows, S. O. (2009). Marital trajectories and mortality among US adults. *American Journal of Epidemiology, 170*(5), 546–555.

Dyrenforth, P. S., Kashy, D. A., Donnellan, M. B., & Lucas, R. E. (2010). Predicting relationship and life satisfaction from personality in nationally representative samples from three countries: The relative importance of actor, partner, and similarity effects. *Journal of Personality and Social Psychology, 99*(4), 690–702.

Engel, G., Olson, K. R., & Patrick, C. (2002). The personality of love: Fundamental motives and traits related to components of love. *Personality and Individual Differences, 32*(5), 839–853.

Etcheverry, P. E., Le, B., Wu, T. F., & Wei, M. (2013). Attachment and the investment model: Predictors of relationship commitment, maintenance, and persistence. *Personal Relationships, 20*(3), 546–567.

Fincham, F. D., & Beach, S. R. H. (1999). Conflict in marriage: Implications for working with couples. *Annual Review of Psychology, 50*, 47–77.

Fraley, R. C., & Shaver, P. R. (2000). Adult romantic attachment: Theoretical developments, emerging controversies, and unanswered questions. *Review of General Psychology, 4*(2), 132–154.

Frech, A., & Williams, K. (2007). Depression and the psychological benefits of entering marriage. *Journal of Health and Social Behavior, 48*(2), 149–163.

Gottman, J. M., & Notarius, C. I. (2002). Marital research in the 20th century and a research agenda for the 21st century. *Family Process, 41*(2), 159–197.

Gove, W. R., & Tudor, J. F. (1973). Adult sex roles and mental illness. *American Journal of Sociology, 78*, 812–835.

Graziano, W. G., Jensen-Campbell, L. A., & Hair, E. C. (1996). Perceiving interpersonal conflict and reacting to it: The case for agreeableness. *Journal of Personality and Social Psychology, 70*(4), 820–835.

Hawkins, D. N., & Booth, A. (2005). Unhappily ever after: Effects of long-term, low-quality marriages on Well-being. *Social Forces, 84*(1), 451–471.

Hazan, C., & Shaver, P. R. (1994). Attachment as an organizational framework for research on close relationships. *Psychological Inquiry, 5*((1)), 1–22.

Horwitz, A. V., White, H. R., & Howell-White, S. (1996). Becoming married and mental health: A longitudinal analysis of a cohort of young adults. *Journal of Marriage and the Family, 58*, 895–907.

Jackson, J. J., Wood, D., Bogg, T., Walton, K., Harms, P., & Roberts, B. W. (2010). What do conscientious people do? Development and validation of the Behavioral Indicators of Conscientiousness Scale (BICS). *Journal of Research in Personality, 44*, 501–511.

Jensen-Campbell, L. A., & Graziano, W. G. (2001). Agreeableness as a moderator of interpersonal conflict. *Journal of Personality, 69*(2), 323–362.

John, O. P., & Srivastava, S. (1999). The Big Five Trait taxonomy: History, measurement, and theoretical perspectives. In L. A. Pervin & O. P. John (Eds.), *Handbook of personality: Theory and research* (pp. 102–138). New York, NY: Guilford Press.

Kenny, D. A. (2018). Reflections on the actor–partner interdependence model. *Personal Relationships, 25*(2), 1–11.

Kenny, D. A., Kashy, D. A., & Cook, W. L. (2006). Methodology in the social sciences (David A. Kenny, Series Editor). *Dyadic data analysis*. New York, NY: Guilford Press.

Kessler, R. C., & Essex, M. (1982). Marital status and depression: The importance of coping resources. *Social Forces, 61*(2), 484–507.

Kiecolt-Glaser, J. K. (2018). Marriage, divorce, and the immune system. *American Psychologist, 73*(9), 1098–1108.

Kiecolt-Glaser, J. K., Loving, T. J., Stowell, J. R., Malarkey, W. B., Lemeshow, S., Dickinson, S. L., & Glaser, R. (2005). Hostile marital interactions, proinflammatory cytokine production, and wound healing. *Archives of General Psychiatry, 62*(12), 1377–1384.

Kiecolt-Glaser, J. K., & Newton, T. L. (2001). Marriage and health: His and hers. *Psychological Bulletin, 127*(4), 472–503.

Kim, E. S., Chopik, W. J., & Smith, J. (2014). Are people healthier if their partners are more optimistic? The dyadic effect of optimism on health among older adults. *Journal of Psychosomatic Research, 76(6), 447–453.*

Kirkpatrick, L. A., & Davis, K. E. (1994). Attachment style, gender, and relationship stability: A longitudinal analysis. *Journal of Personality and Social Psychology, 66*(3), 502–512.

Kurdek, L. A. (1993). Predicting marital dissolution: A 5-year prospective longitudinal study of newlywed couples. *Journal of Personality and Social Psychology, 64*(2), 221–242.

Lavy, S., Mikulincer, M., & Shaver, P. R. (2013). Intrusiveness from an attachment theory perspective: A dyadic diary study. *Personality and Individual Differences, 55*(8), 972–977.

Ledermann, T., & Kenny, D. A. (2017). Analyzing dyadic data with multilevel modeling versus structural equation modeling: A tale of two methods. *Journal of Family Psychology, 31*(4), 442–452.

Lee-Baggley, D., Preece, M., & DeLongis, A. (2005). Coping with interpersonal stress: Role of big five traits. *Journal of Personality, 73*(5), 1141–1180.

Malouff, J. M., Thorsteinsson, E. B., Schutte, N. S., Bhullar, N., & Rooke, S. E. (2010). The Five-Factor Model of personality and relationship satisfaction of intimate partners: A meta-analysis. *Journal of Research in Personality, 44*(1), 124–127.

Mikulincer, M., & Shaver, P. R. (2012). Attachment theory expanded: A behavioral systems approach. In K. Deaux & M. Snyder (Eds.), *Oxford Library of Psychology. The Oxford handbook of personality and social psychology* (pp. 467–492). New York, NY: Oxford University Press.

Mikulincer, M., & Shaver, P. R. (2016). *Attachment in adulthood: Structure, dynamics and change* (2nd ed.). New York, NY: Guilfold Press.

Miller, G. E., Dopp, J. M., Myers, H. F., Stevens, S. Y., & Fahey, J. L. (1999). Psychosocial predictors of natural killer cell mobilization during marital conflict. *Health Psychology, 18*(3), 262–271.

Murray, S. L., Rose, P., Bellavia, G. M., Holmes, J. G., & Kusche, A. G. (2002). When rejection stings: How self-esteem constrains relationship-enhancement processes. *Journal of Personality and Social Psychology, 83*(3), 556–573.

Neff, L. A., & Karney, B. R. (2009). Stress and reactivity to daily relationship experiences: How stress hinders adaptive processes in marriage. *Journal of Personality and Social Psychology, 97*(3), 435–450.

Nilsson, P. M., Engstrom, G., & Hedblad, B. (2008). Long-term predictors of increased mortality risk in screened men with new hypertension: The Malmo preventive project. *Journal of Hypertension, 26*(12), 2284–2294.

Noftle, E. E., & Shaver, P. R. (2006). Attachment dimensions and the big five personality traits: Associations and comparative ability to predict relationship quality. *Journal of Research in Personality, 40*(2), 179–208.

Overbeek, G., Vollebergh, W., de Graaf, R., Scholte, R., de Kemp, R., & Engels, R. (2006). Longitudinal associations of marital quality and marital dissolution with the incidence of DSM-III-R disorders. *Journal of Family Psychology, 20*(2), 284–291.

Pietromonaco, P. R., Uchino, B., & Dunkel Schetter, C. (2013). Close relationship processes and health: Implications of attachment theory for health and disease. *Health Psychology, 32*(5), 499–513.

Powers, S. I., Pietromonaco, P. R., Gunlicks, M., & Sayer, A. (2006). Dating couples' attachment styles and patterns of cortisol reactivity and recovery in response to a relationship conflict. *Journal of Personality and Social Psychology, 90*(4), 613–628.

Proulx, C. M., Helms, H. M., & Buehler, C. (2007). Marital quality and personal well-being: A meta-analysis. *Journal of Marriage and Family, 69*(3), 576–593.

Roberts, B. W., Kuncel, N. R., Shiner, R., Caspi, A., & Goldberg, L. R. (2007). The power of personality: The comparative validity of personality traits, socioeconomic status, and cognitive ability for predicting important life outcomes. *Perspectives on Psychological Science, 2*(4), 313–345.

Roberts, B. W., Jackson, J. J., Fayard, J. V., Edmonds, G., & Meints, J. (2009a). Conscientiousness. In M. R. Leary & R. H. Hoyle (Eds.), *Handbook of individual differences in social behavior* (pp. 369–381). New York, NY: The Guilford Press.

Roberts, B. W., Smith, J., Jackson, J. J., & Edmonds, G. (2009b). Compensatory conscientiousness and health in older couples. *Psychological Science, 20*(5), 553–559.

Robins, R. W., Caspi, A., & Moffitt, T. E. (2002). It's not just who you're with, it's who you are: Personality and relationship experiences across multiple relationships. *Journal of Personality, 70*(6), 925–964.

Robles, T. F., & Kiecolt-Glaser, J. K. (2003). The physiology of marriage: Pathways to health. *Physiology & Behavior, 79*(3), 409–416.

Robles, T. F., Slatcher, R. B., Trombello, J. M., & McGinn, M. M. (2014). Marital quality and health: A meta-analytic review. *Psychological Bulletin, 140*(1), 140–187.

Rogers, R. G. (1995). Marriage, sex, and mortality. *Journal of Marriage and the Family, 57*(2), 515–526.

Rohrer, J. E., Bernard, M. E., Zhang, Y., Rasmussen, N. H., & Woroncow, H. (2008). Marital status, feeling depressed and self-rated health in rural female primary care patients. *Journal of Evaluation in Clinical Practice, 14*(2), 214–217.

Rook, K. S., & Zettel, L. A. (2005). The purported benefits of marriage viewed through the lens of physical health. *Psychological Inquiry, 16*(2), 116–121.

Russell, R. J. H., & Wells, P. A. (1994). Predictors of happiness in married couples. *Personality and Individual Differences, 17*(3), 313–321.

Shaver, P. R., & Mikulincer, M. (2007). Adult attachment strategies and the regulation of emotion. In J. J. Gross (Ed.), *Handbook of emotion regulation* (pp. 446–465). New York, NY: The Guilford Press.

Simpson, J. A., Rholes, W. S., & Nelligan, J. S. (1992). Support seeking and support giving within couples in an anxiety-provoking situation: The role of attachment styles. *Journal of Personality and Social Psychology, 62*(3), 434–446.

Slatcher, R. B., & Selcuk, E. (2017). A social psychological perspective on the links between close relationships and health. *Current Directions in Psychological Science, 26*(1), 16–21.

Solomon, B. C., & Jackson, J. J. (2014). Why do personality traits predict divorce? Multiple pathways through satisfaction. *Journal of Personality and Social Psychology, 106*(6), 978–996.

Statistics Canada. (2015). *Leading causes of death in Canada*, 2009. Retrieved from https://www150.statcan.gc.ca/n1/pub/84-215-x/2012001/hl-fs-eng.htm

Taraban, C. B., Hendrick, S. S., & Hendrick, C. (1998). Loving and liking. In P. A. Andersen & L. K. Guerrero (Eds.), *Handbook of communication and emotion* (pp. 331–351). San Diego, CA: Academic.

Tobin, R. M., Graziano, W. G., Vanman, E. J., & Tassinary, L. G. (2000). Personality, emotional experience, and efforts to control emotions. *Journal of Personality and Social Psychology, 79*(4), 656–669.

Tucker, J. S., Friedman, H. S., Wingard, D. L., & Schwartz, J. E. (1996). Marital history at midlife as a predictor of longevity: Alternative explanations to the protective effect of marriage. *Health Psychology, 15*(2), 94–101.

Umberson, D. (1992). Gender, marital status and the social control of health behavior. *Social Science & Medicine, 34*(8), 907–917.

Umberson, D., & Williams, K. (1999). Family status and mental health. In C. S. Aneshensel & J. C. Phelan (Eds.), *Handbook of sociology of mental health* (pp. 225–253). Dordrecht, The Netherlands: Kluwer Academic Publishers.

Umberson, D., Williams, K., Powers, D. A., Liu, H., & Needham, B. (2006). You make me sick: Marital quality and health over the life course. *Journal of Health and Social Behaviour, 47*(1), 1–16.

Wadworth, T. (2016). Marriage and subjective well-being: How and why context matters. *Social Indicators Research, 126*(3), 1025–1048.

Waite, L. J., & Gallagher, M. (2000). *The case for marriage: Why married people are happier, healthier and better off financially*. New York, NY: Doubleday.

Watson, D., Hubbard, B., & Wiese, D. (2000). Self–other agreement in personality and affectivity: The role of acquaintanceship, trait visibility, and assumed similarity. *Journal of Personality and Social Psychology, 78*(3), 546–558.

Whisman, M. A. (2001). The association between depression and marital dissatisfaction. In S. R. H. Beach (Ed.), *Marital and family processes in depression: A scientific foundation for clinical practice* (pp. 3–24). Washington, DC: American Psychological Association.

Wickrama, K. A. S., Lorenz, F. O., Conger, R. D., & Elder, G. H. (1997). Marital quality and physical illness: A latent growth curve analysis. *Journal of Marriage and Family, 59*, 143–155.

Williams, K., & Umberson, D. (2004). Marital status, marital transitions, and health: A gendered life course perspective. *Journal of Health and Social Behaviour, 45*(1), 81–98.

Chapter 3
Aging with Purpose: Developmental Changes and Benefits of Purpose in Life Throughout the Lifespan

Gabrielle N. Pfund and Nathan A. Lewis

3.1 Purpose as a Contributor and Correlate for Healthy Aging Across the Lifespan

Purpose is a dynamic construct that fluctuates in prevalence throughout the lifespan, but consistently predicts desirable outcomes regardless of one's age. *Sense of purpose* can be understood as the extent to which one feels that they have personally meaningful goals and directions guiding them through life (Ryff, 1989), while a *purpose in life* refers to those specific goals and directions (McKnight & Kashdan, 2009). Purpose is not viewed as simply a goal one must achieve before moving onto another, but an overarching theme that penetrates the various smaller obtainable goals that one pursues throughout their lives (McKnight & Kashdan, 2009; Hill, Burrow, Brandenberger, et al., 2010). As described by Damon, Menon, and Bronk (2003), one's purpose in life should, in fact, not be easy to attain, if attainable at all. By this definition, one's purpose in life would not solely be to have a child or to feed the homeless at a soup kitchen. One's purpose in life can be understood as more of a general orientation, like to spend time with and care for one's family, or to help those in need. By considering purpose through this perspective, purpose reflects something that one is able to constantly pursue and take esteemed steps toward achieving, without being in a position of completing a distinct objective and being left purposeless.

G. N. Pfund (✉)
Department of Psychological & Brain Sciences, Washington University in St. Louis, St. Louis, MO, USA
e-mail: gabrielle.pfund@wustl.edu

N. A. Lewis
Department of Psychology, University of Victoria, Victoria, BC, Canada

Institute on Aging and Lifelong Health, University of Victoria, Victoria, BC, Canada

P. L. Hill, M. Allemand (eds.), *Personality and Healthy Aging in Adulthood*, International Perspectives on Aging 26,
https://doi.org/10.1007/978-3-030-32053-9_3

Much like the Big Five personality traits discussed throughout this volume, sense of purpose in life shows relative stability over time, and exhibits discernible developmental trends across the lifespan. For instance, adolescence is typically marked by the genesis of the purpose exploration process, where individuals try to find their purpose in the midst of trying to find themselves and their place in the world (Bronk, Hill, Lapsley, Talib, & Finch, 2009; Burrow, O'Dell, & Hill, 2010). Emerging adulthood is a continuation of this process, with some having already identified their purpose while others continue their search (Bronk et al., 2009; Sumner, Burrow, & Hill, 2015). As people transition into middle adulthood, most have committed to their purpose in life, creating a stabilization of a relatively higher sense of purpose during this period (e.g., Ko, Hooker, Geldhof, & McAdams, 2016; Pinquart, 2002). However, when entering older adulthood, many report a lower sense of purpose than their middle adult peers (Karasawa et al., 2011). As situations change in the midst of retirement, death of loved ones, and diminution of physical and mental abilities, many individuals experience a decrease in sense of purpose or the loss of their purpose in life altogether.

While the presence and prevalence of purpose throughout the lifespan changes, its inherent value remains consistent. Regardless of where one is developmentally, higher levels of purpose consistently predict desirable mental, emotional, and physical health outcomes. Sense of purpose can foster benefits concurrently, meaning that individuals with a higher sense of purpose also tend to experience more desirable outcomes in the midst of their developmental period. However, the extent of purpose's value is also forward reaching, insofar as it may foster positive transitions throughout the lifespan developmental process. Throughout this chapter, we will review developmental changes in sense of purpose across the lifespan and highlight current research underscoring the benefits of purposefulness on health and well-being. Furthermore, we will discuss the mechanisms that shape these associations between purpose and positive health outcomes, as well as potential methods to harness a purpose for those who have yet to develop or have lost one. We will begin by discussing the purpose development process, typically experienced during adolescence and emerging adulthood, and the associated identity development and well-being outcomes. From there, we will discuss the developmental trends of purpose within the context of middle and older adulthood, and the cognitive and physical health outcomes with which it is concurrently and longitudinally associated. We will conclude this chapter by discussing potential pathways to assess mechanisms and correlates of purpose, as well as to introduce new research questions in the purpose and healthy aging literature.

3.2 Developmental Trends in Purpose During Adolescence and Emerging Adulthood

Adolescence is a delicate time in individuals' lives where they experience and maneuver their way through difficult developmental processes. Identity development tends to occur in earnest at this time, during which an adolescent tries to understand and figure out who they are (Kroger & Marcia, 2001). This process is understood as a time in which individuals explore and commit to various components of their identities, whether *personal identities*, in which a person determines the aspects of themselves that are unique from others, or *social identities*, in which a person determines their association with particular in-groups (Brewer & Gardner, 1996; Sim, Goyle, McKedy, Eidelman, & Correll, 2014). During this process, individuals both *explore* potential avenues that encompass who they are before they *commit* to aspects that best describes their identities (Kroger & Marcia, 2001). The exploration and solidification of one's self-concept and identity is a prime component of adolescence.

Just as one works through the identity development process, so does one work through the purpose development process. Previous research has used identity development frameworks to understand this purpose process, marked by *purpose exploration*, searching for one's purpose, and leading to *purpose commitment*, or identifying and committing to one's general purpose in life (Bronk et al., 2009; Sumner et al., 2015). Research has found purpose exploration and commitment to be closely related to the processes of identity exploration and commitment among adolescents, with purpose exploration being negatively related to identity commitment (Burrow et al., 2010; Hill & Burrow, 2012). Moreover, when following individuals over time, adolescents who increased on identity commitment also reported an increase on purpose commitment (Hill & Burrow, 2012). However, change in purpose commitment was positively associated with change in purpose exploration. These findings lend themselves to an important developmental story: As one begins to narrow in on and commit to a purpose in life, they may also begin to explore the depth and various pathways that this purpose in life may provide.

In addition to considering the trajectories and implications of purpose in adolescence, researchers have also investigated whether adolescents are able to grasp the construct of purpose itself, and if their conceptualization of purpose aligns with the operationalization set forth by researchers. In order to address this concern, one study asked both public and private high school students to define what purpose in life means (Hill, Burrow, O'Dell, & Thornton, 2010). Participants' responses were coded to determine whether they reflected five purpose orientations common in the purpose in life literature: foundation and direction, happiness, prosociality, religiosity, and occupational/financial goals. Nearly all participants (98%) mentioned at least one of these orientation themes in defining purpose in life, with having a foundation or direction being the most prevalent theme in these high school students' definitions (82%). In fact, about 70% of participants mentioned multiple themes in their definition of purpose, illustrating their understanding that purpose is not necessarily a one-size-fits-all construct. This study illustrated that purpose is both understood by and important to adolescents, building upon previous research suggesting that adolescence marks the beginning of many individuals' search for purpose (Bronk et al., 2009; Burrow et al., 2010).

The process of exploring and identifying one's purpose in life continues from adolescence into emerging adulthood. While adolescence is generally understood as occurring between the ages of 12 to 17 and emerging adulthood the ages of 18 to 25 or 30 (Arnett, 2000), there is fluidity in these developmental experiences. Because of these blurred categorizations, previous research evaluating purpose trajectories across these age ranges has taken a couple different approaches, sometimes considering these ages more holistically while other times separating them into "adolescence" and "emerging adulthood" more distinctly. With this in mind, we will discuss the benefits of purpose during these periods, while noting similarities and distinctions for each when appropriate.

3.3 Purpose Promotes Adaptive Development during Adolescence and Emerging Adulthood

Purpose has been deemed an important tool and asset that aids adolescents in navigating positive development (Bronk, 2013; Burrow et al., 2010; Damon et al., 2003). The benefits of purpose in adolescence and emerging adulthood are generally demonstrated with respect to psychological outcomes, with purpose aiding in the promotion of positive well-being (Bronk et al., 2009; Mariano & Savage, 2009). Positive correlates of purpose during this time period are identity development, life satisfaction, subjective well-being, and general affect (Bronk, 2012; Bronk et al., 2009; Burrow et al., 2010; Burrow & Hill, 2011). Furthermore, there are adaptive dispositions associated with purpose, like conscientiousness and grit (Hill & Burrow, 2012; Malin, Liauw, & Damon, 2017), that can aid an adolescent or emerging adult in attaining or maintaining these desirable states (Hill, Edmonds, Peterson, Luyckx, & Andrews, 2016).

The importance of these developmental processes goes beyond being able to claim one knows who they are and what their purpose is—the differences in these exploration and commitment outcomes have critical implications for the individuals in each stage. For example, findings regarding the connection between purpose exploration and well-being have presented a mixed picture thus far. When considering purpose exploration, some research has found that searching for a purpose is associated with higher life satisfaction in both adolescents and emerging adults (Bronk et al., 2009). Other studiess have found that purpose exploration had no relation to well-being outcomes such as positive affect (Burrow et al., 2010), and actually predicted lower life satisfaction and higher negative affect in emerging adults (Sumner et al., 2015).

Regarding purpose commitment in adolescence and emerging adulthood, the findings are more consistent and much more positive. Purpose commitment is positively associated with life satisfaction, positive affect, and happiness in adolescents (Bronk et al., 2009; Burrow et al., 2010; Burrow & Hill, 2011). Furthermore, while identity commitment is also related to various well-being outcomes, purpose has been shown to fully mediate the relationships between identity commitment and well-being aspects, such as positive and negative affect (Burrow & Hill, 2011). Purpose also has been described as an important component in initiating the transition from adolescence to emerging adulthood (Hill & Burrow, 2012), as well as a motivator of positive transitioning between these stages (Hill, Burrow, Brandenberger, Lapsley, & Quaranto, 2010). Some have even found that purpose is connected to the desirable personality traits that aid in optimal youth development, such as generosity, gratitude, and compassion (Malin et al., 2017; Mariano & Savage, 2009) as well as mature development such as conscientiousness (Hill & Burrow, 2012). While purpose commitment and identity commitment are intertwined processes, purpose commitment is the powerful promoter for adolescent psychological health.

Purpose commitment also has desirable outcomes for emerging adults, with people experiencing greater life satisfaction, well-being, and positive affect, as well as lower levels of negative affect (Bronk et al., 2009; Hill, Burrow, & Bronk, 2016; Hill, Sumner, & Burrow, 2014). Once again, emerging adulthood follows similar patterns as adolescence, with purpose completely mediating the relationship between identity commitment and psychological well-being (Burrow & Hill, 2011). Purpose commitment is a significant contributor to greater life satisfaction, higher positive affect, and lower negative affect, when compared to markers of identity commitment and exploration of both identity and purpose (Sumner et al., 2015). In other words, being committed to one's purpose in life has been found to be a greater contributor to well-being than identity commitment. Purpose has also been shown to predict well-being in emerging adulthood even when controlling for covariates such as the Big Five personality traits (Hill, Edmonds, Peterson, et al., 2016). Even after adjusting for other powerful psychological health contributors, purpose continues to show its relevance and prowess as a contributor to adolescents and emerging adults' mental health.

In addition to promoting positive outcomes such as psychological well-being, purpose is also associated with a number of characteristics that may in turn facilitate positive youth development. For example, both adolescents and emerging adults who report having a purpose also report higher levels of hope (Bronk et al., 2009; Burrow et al., 2010), defined as a combination of *pathways*, which is the belief that there is a means to reach one's goal, and *agency*, which is the belief that one has the necessary motivation to utilize those pathways to reach one's goal (Snyder, Rand, & Sigmon, 2005). Furthermore, when university students reported a higher sense of purpose and positive affect at the beginning of the semester, those who reported higher initial purpose levels were more likely to gain on grit over the semester (Hill, Burrow, & Bronk, 2016). These associations may help elucidate mechanisms that may partly explain the consistent connections between purpose and psychological health for adolescents and emerging adults.

Though the purpose research in these age groups focuses on outcomes associated with optimal youth development and psychological well-being, the implications of purpose development during this time may extend throughout the lifespan. Indeed, finding a purpose early on can foretell later psychological, physical, and cognitive health benefits. Next, we will evaluate the implications of purpose later in the lifespan.

3.4 Developmental Trajectories of Sense of Purpose in Middle and Older Adulthood

Whereas adolescence and emerging adulthood are viewed as periods of fluidity in purpose exploration, middle adulthood is typically marked by greater stability in sense of purpose over time. However, several studies have suggested that sense of purpose in life begins to decline in the latter half of the lifespan, with older adults showing diminished purposefulness compared to younger counterparts. A meta-analysis of over 70 studies on sense of purpose in middle and older adulthood reported a small age-related decline in sense of purpose beginning in midlife, with these changes being more pronounced following retirement age (Pinquart, 2002). One potential explanation for this trend is that sense of purpose may be closely tied to our major life roles, which may decrease with advancing age and after retirement. Older individuals often face a number of unique challenges to their purpose such as retirement, declining health or functional ability, widowhood, and changes in social structure. Each of these events has the potential to influence the type of goal pursuits in which an individual engages, and therefore, the loss of these could lead to a less salient sense of purpose in older individuals.

However, a major limitation of this area has been the reliance on cross-sectional data. The studies synthesized in the above meta-analysis each report mean-level differences between younger and older age groups, but do not reflect how sense of purpose may change within individuals over time. This is concerning as cross-sectional comparisons across age-heterogeneous samples may greatly inflate age-related effects and do not account for the influence of between-cohort differences (see Hofer & Sliwinski, 2001). As such, longitudinal designs are needed to make claims about within-person changes in purpose over time. More recent research using longitudinal data has found that sense of purpose does decline with advancing age, though this is not the case for all individuals. Using two large longitudinal studies, Springer, Pudrovska, and Hauser (2011) examined change in sense of purpose across a 10-year interval in young, middle, and older adult age cohorts. They observed a small decline in sense of purpose between assessments, particularly among the oldest age group. Similarly, Hill and colleagues (2015) assessed within-person changes in sense of purpose among older men across a 3-year span but found little mean-level change over time. Finally, one study found that retired individuals in the Health and Retirement Study experienced a mean-level decline in sense of purpose over an 8 year period, though there was significant individual-level variability in purpose change over time (Hill & Weston, 2019). An important finding in these studies is that there is considerable heterogeneity in the changes in sense of purpose of adults over time. In other words, though purposefulness may show slight mean-level declines approaching older adulthood, not everyone follows this trend—many individuals show relative stability or even growth in purposefulness over time.

Given the differences in individual trajectories of change in sense of purpose, an emerging literature has sought to examine factors that may influence how individuals change in purpose over time. One such precipitating factor may be physical health, as many studies have reported that better health and physiological functioning are associated with higher sense of purpose (see Pinquart, 2002). Declining health may limit the ability to engage in goals central to a purpose, causing individuals to feel helpless and directionless. This may be particularly true for ailments with a sudden onset or those leading to disability. For example, one study of longitudinal changes following stroke onset observed declines in sense of purpose over time relative to pre-stroke purpose levels, with diminished physical and cognitive functioning predicting a lower sense of purpose (Lewis, Brazeau, & Hill, 2018). However, this was not the case for individuals who had suffered a stroke several years prior to baseline, suggesting that after a period of time some patients return to relatively stable levels of purposefulness. Though events such as retirement or the onset of a health condition may disrupt past roles contributing to one's sense of purpose, older individuals may adapt to derive purpose from other pursuits.

In the face of accumulating age-related losses, individuals may engage in compensatory strategies to optimize their functioning and continue to pursue similar goals (Baltes, Staudinger, & Lindenberger, 1999). Recent research following adults around the transition into retirement found that frequent engagement in a number of physical, social, cognitive, and prosocial leisure activities mitigated against decline in purposefulness over an 8-year span (Lewis & Hill, 2019). Further, the benefits of leisure activity engagement were more pronounced among newly retired individuals, suggesting that living an engaged lifestyle may help individuals compensate for loss of roles in other areas of life. Similarly, older adults may rely on social support to pursue purposeful aims despite accruing role losses. The Pinquart (2002) meta-analysis found that social factors such as social network size and relationship quality were among the strongest predictors of sense of purpose in middle-to-older adulthood. In addition, longitudinal research in individuals around the transition into older adulthood has found that perceived social support from one's spouse, children, and friends helps to buffer against decline in purpose over an 8-year period (Weston, Lewis, & Hill, 2019). As such, though several studies have observed age-associated changes in sense of purpose, not all individuals are destined to decline in purposefulness as they age. Factors such as activity engagement and social support might help individuals preserve sense of purpose, allowing them to continue to garner benefits from purpose into older adulthood.

3.5 Purpose Promotes Desirable Health Outcomes throughout Adulthood

A myriad of research has highlighted the value of having a sense of purpose in life in fostering positive health outcomes in middle and older adulthood. Purpose is associated with reduced risk of physical disability (Boyle, Buchman, Barnes, & Bennett, 2010), cardiovascular conditions such as stroke and myocardial infarction (Kim, Sun, Park, Kubzansky, & Peterson, 2013; Kim, Sun, Park, & Peterson, 2013), Alzheimer's disease and mild cognitive impairment (Boyle et al., 2010; Boyle et al., 2012), and mortality (Boyle, Barnes, Buchman, & Bennett, 2009; Hill & Turiano, 2014). Another recent study showed that purpose in life is associated with lower levels of glycosylated hemoglobin, a marker of long-term glucose regulation associated with elevated risk for Type-II diabetes, cardiovascular disease, and mortality (Boylan, Tsenkova, Miyamoto, & Ryff, 2017). One proposed mechanism for the association between sense of purpose in life and positive health outcomes is that purposeful individuals may be more likely to pursue a number of health-protective behaviors. For instance, individuals higher in sense of purpose report more frequent engagement in moderate and vigorous physical activities (Hill, Edmonds, & Hampson, 2017; Holahan, Holahan, & Suzuki, 2008). This research is supported by recent findings linking higher sense of purpose with accelerometer-measured physical activity and general movement throughout the day (Hooker & Masters, 2016). Additional associations have also been observed for vegetable intake, sleep quality, and dental care (Hill et al., 2017; Kim, Hershner, & Strecher, 2015). Purpose also appears to predict health behaviors in the context of health care utilization and treatment adherence. Purposeful individuals are more likely to utilize preventative medical screenings such as cholesterol tests, mammograms, pap smears, and prostate exams (Kim, Strecher, & Ryff, 2014). Together, these findings point to purposeful adults taking a more active approach to their health through a number of health-promoting behaviors.

Another avenue of research involves the association of sense of purpose with coping and physiological reactivity to stressful experiences. Several studies have found that individuals higher in purpose report more positive reactions in response to stressors, including fewer depressive symptoms and less engagement in avoidant coping strategies (e.g., Wang, Lightsey, Pietruszka, Uruk, & Wells, 2007). Additional findings have documented a relationship between sense of purpose in life and more objective measures of stress response. For example, one study showed that purposeful individuals experience reduced heart rate variability after being presented with a stress-inducing video, with those higher in purpose reporting less anxiety and a more rapid return to resting heart rate (Ishida, 2006). A related study examined emotion recovery in adults presented with negative picture stimuli through eye blink response, a measure sensitive to emotional state in response to stressors (Schaefer et al., 2013). Similar to the previous findings, those higher in sense of purpose exhibited better emotional recovery after viewing the negative stimulus. Moreover, those higher in purpose have shown to have consistently lower salivary levels of the stress hormone cortisol throughout the day (Ryff et al., 2006). Taken together, these findings suggest a healthier stress response profile: when stressors are no longer present, purposeful individuals are able to quickly reduce autonomic activation and return to a resting state.

Extending beyond health behaviors and stress reactivity, sense of purpose is associated with healthier biological risk profiles assessed via several inflammatory and functional biomarkers. Sense of purpose is related to a healthier cardiovascular risk profile, including lower levels of inflammatory cytokines, higher levels of high-density lipoprotein ("good cholesterol"), and down-regulation of pro-inflammatory genes (Fredrickson et al., 2015; Ryff et al., 2006; Zilioli, Slatcher, Ong, & Gruenewald, 2015). Additionally, individuals reporting a higher sense of purpose have been found to have significantly lower levels of soluble interleukin-6 receptors, a cytokine receptor which serves to amplify the inflammatory response (Friedman, Hayney, Love, Singer, & Ryff, 2007). One recent study by Zilioli et al. (2015) showed that sense of purpose predicted reduced allostatic load—a composite measure of several biomarkers representing functioning across a number of physiological systems. This index included markers of lipid metabolism such as body mass index and cholesterol levels, inflammatory cytokines such as interleukin-6 (IL-6) and C-reactive protein, and parasympathetic nervous system activity. Therefore, the finding of decreased allostatic load in purposeful adults reflects reduced physiological strain on immune, cardiovascular, and other bodily systems, suggesting that these individuals may be less prone to diseases resulting from chronic wear on these systems. Beyond these biomarkers, purpose also is longitudinally related to better grip strength and faster walking speed (Kim, Kawachi, Chen, & Kubzansky, 2017), two functional biomarkers known to predict cognitive change and dementia risk in late life (MacDonald et al., 2017).

Further underscoring its benefits for adult development, a growing literature has linked purpose in life with greater cognitive health. Middle and older adults reporting a higher sense of purpose perform better on several cognitive assessments including measures of executive functioning, memory, and processing speed (Lewis, Turiano, Payne, & Hill, 2017; Windsor, Curtis, & Luszcz, 2015). A number of potential mechanisms may account for these between-person differences in cognitive functioning. Higher sense of purpose is moderately associated with educational attainment and socioeconomic factors that may support cognitive ability (Hill, Turiano, Mroczek, & Burrow, 2016; Pinquart, 2002). Further, purposeful individuals may benefit from increased engagement in complex cognitive tasks associated with purpose-driven goals, leading to improved functioning through cognitive enrichment (e.g., Hertzog, Kramer, Wilson, & Lindenberger, 2008). Considering non-normative cognitive development, prospective studies have found that more purposeful individuals have a reduced risk for developing mild cognitive impairment, Alzheimer's Disease, and other forms of dementia (Boyle et al., 2010, 2012). Sense of purpose may reduce risk for cognitive impairment by contributing to the cognitive reserve capacity of older adults. Cognitive reserve reflects the brain's ability to compensate for pathological changes such as brain injury or neurodegeneration, allowing some individuals to maintain cognitive functions even when widespread pathology is present (Stern, 2002). Research by Boyle et al. (2012) examined longitudinal changes in sense of purpose and cognitive functioning in older adults who also underwent brain autopsy upon death. They found that even after adjusting for markers of reserve such as education, purpose in life moderated the association between Alzheimer's pathology (amyloid plaques and neurofibrillary tangles) and cognition. In other words, more purposeful individuals exhibited higher cognitive functioning even when they displayed the neurological hallmarks of the disease. Though it is not clear why purpose is associated with cognitive reserve, the authors suggest that having a strong sense of purpose may promote goal pursuit and engagement in mentally stimulating activities.

3.6 Conclusion and Future Directions

In sum, purpose in life appears to contribute to improved health outcomes across multiple domains, implying that purpose works through a multitude of physical, psychological, and behavioral influences to promote health (see Fig. 3.1). Moreover, purpose has been shown to predict positive health outcomes even after adjusting for factors such as personality, psychological well-being, and demographic factors (e.g., Boyle et al., 2010; Hill & Turiano, 2014). From adolescence to older adulthood,

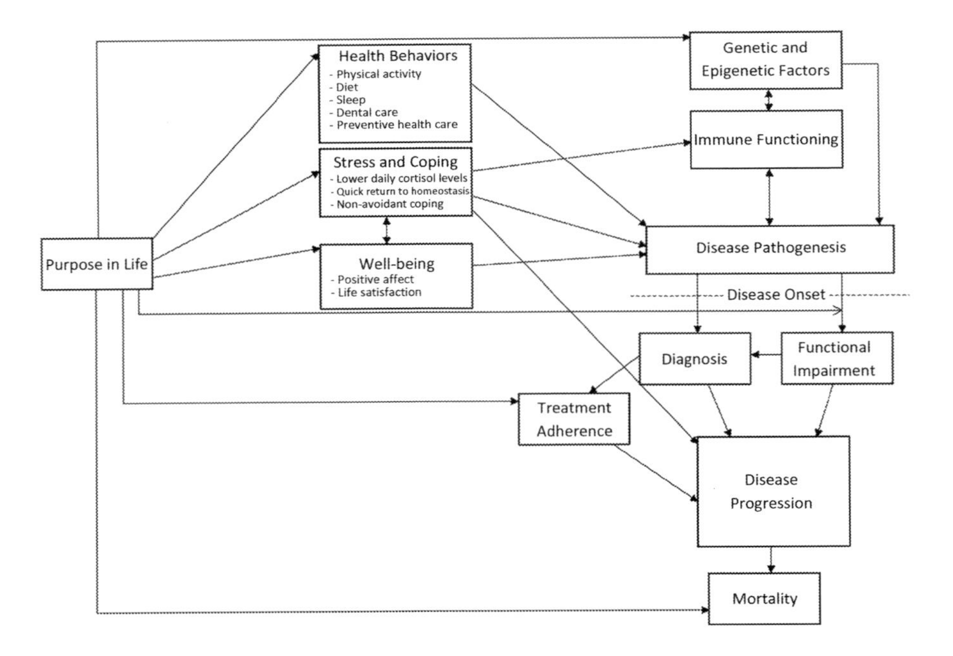

Fig. 3.1 Theoretical and empirical foundations for the role of purpose on physical, psychological, and behavioral outcomes

purpose predicts desirable outcomes for those who have it. That said, there are still many opportunities to both understand the how and why of purpose.

3.6.1 Future Directions for Research on Adolescents and Emerging Adults

The majority of the adolescent and emerging adulthood literature has focused on the psychological and developmental benefits of purpose, such as life satisfaction and identity development. While these findings are important to understand well-being and growth within these individuals, there is room to expand upon various research questions that are prevalent in purpose research on middle and older adults. For example, sense of purpose is positively associated with both better health behaviors and better health outcomes in older adults. Understanding that purpose is a lifespan developmental process, perhaps evaluating these same outcomes earlier on in the lifespan will further elucidate the mechanisms underlying associations between sense of purpose and health outcomes later. Furthermore, those who commit to a purpose in life earlier may experience better health outcomes later in the lifespan than those who commit to a purpose in life later on, given that they are more likely to engage in positive health behavior habits at an earlier age. If purpose promotes valuable health behaviors throughout the lifespan, it then becomes even more important to understand the strength of these associations when individuals are able to commit to a purpose in life at an earlier age.

Furthermore, though a variety of well-being outcomes have been explored in the context of sense of purpose for these ages, the mechanisms behind these relationships still remain unclear. It is suggested that purpose is related to but distinct from well-being (Pfund & Hill, 2018), though the nature of this association could be valuably investigated through potential mediators such as hope and grit. Perhaps those who experience a higher sense of purpose are able to maintain greater well-being due to a better handling of obstacles that appear in their path. While a large section of purpose research has focused on adolescents and emerging adults, there are still many opportunities with depth and breadth to explore.

3.6.2 Future Directions for Research on Middle and Older Adults

Another future research avenue is evaluating how the content of one's purpose in life may predict one's ability to maintain purposefulness and avoid losing a sense of purpose with the transitions that many face in older adulthood. For instance, those whose purpose in life is more rooted in their work may be more likely to experience a greater decrease in sense of purpose following retirement, while someone whose purpose in life is oriented more prosocially will be able to continue pursuing their purpose outside of their work so they are able to maintain a higher sense of purpose in the midst of these challenging transitions. With the potential loss of one's purpose in older adulthood, it is important to understand why some experience declining purposefulness as well as identify ways to help those who experience a loss find a purpose again.

Further, most of the previous research on purpose in adulthood has focused on mean-level trends over time (such as stability in midlife, decline in older adulthood) despite accruing research pointing to significant heterogeneity in longitudinal purpose change (e.g., Hill & Turiano, 2014; Hill & Weston, 2019). More work is needed to address individual differences in purpose change over time and identify antecedents of these changes. For example, techniques such as growth mixture modeling or latent class analysis (see Jung & Wickrama, 2008) offer promising avenues for exploring subpopulations with distinct purpose trajectories over time and would allow researchers to examine unique characteristics of these groups. Methods such as these could help to clarify why some individuals experience relative stability or even increases in sense of purpose leading into older adulthood, whereas others decline.

3.6.3 Future Directions Throughout the Lifespan

Considering the far reach of purpose's power, there is one final prominent theme of research that is necessary in understanding purpose. Though purpose has shown itself to be an important factor throughout the developmental lifespan, additional research is needed to investigate how to help individuals who have yet to find a purpose and how to aid in the rediscovery of purpose for those who have lost theirs. Moving forward, research should develop interventions and various purpose enhancing strategies to promote development and maintenance of a sense of purpose. Previous research has theorized three main pathways that may lead to a purpose in life: proactive engagement, social learning, and reactive development (Kashdan & McKnight, 2009). While none of these pathways need to be taken exclusively, future research could investigate both the efficiency and effectiveness of each pathway individually, as well as ways to promote specific pathways to purpose.

The first pathway to purpose is considered "proactive" because an individual seeks out various opportunities and avenues that will help them then narrow in on committing to a purpose in life. The second route, social learning, could lead to purpose commitment through mentoring opportunities that could enlighten individuals on goals accessible to them, as well as being surrounded by purposeful people that help one understand the power of purpose engagement. The third pathway, labeled the reactive route, refers to developing a purpose in life following the occurrence of an event, so a purpose that is spurred from a reaction. This purpose could take form in a variety of ways, such as experiencing a traumatic event and wanting to help others who have faced similar experiences. Each of these pathways to purpose may look different depending on where one is in the lifespan. Those who are younger may be exploring opportunities unique to them while those who are older may have experienced purpose derailment (loss of their previously committed purpose) could be experiencing a reinvigoration of such from experiences of the past. Some potential opportunities to evaluate the effectiveness of these pathways in committing to a purpose in life would be to assess participation in community service activities for proactive engagement, mentorship support for social learning, and reflection on previous meaningful events for reactive development. These are a few of the numerous opportunities to understand how to help encourage individuals to both explore and later commit to their purpose in life, so they, too, can experience the variety of emotional, physical, and cognitive benefits associated with purposeful aging.

References

Arnett, J. J. (2000). Emerging adulthood: A theory of development from the late teens through the twenties. *American Psychologist, 55*(5), 469.

Baltes, P. B., Staudinger, U. M., & Lindenberger, U. (1999). Lifespan psychology: Theory and application to intellectual functioning. *Annual Review of Psychology, 50*(1), 471–507.

Boylan, J. M., Tsenkova, V. K., Miyamoto, Y., & Ryff, C. D. (2017). Psychological resources and glucoregulation in Japanese adults: Findings from MIDJA. *Health Psychology, 36*(5), 449.

Boyle, P. A., Barnes, L. L., Buchman, A. S., & Bennett, D. A. (2009). Purpose in life is associated with mortality among community-dwelling older persons. *Psychosomatic Medicine, 71*(5), 574–579.

Boyle, P. A., Buchman, A. S., Barnes, L. L., & Bennett, D. A. (2010). Effect of a purpose in life on risk of incident alzheimer disease and mild cognitive impairment in community-dwelling older persons. *Archives of General Psychiatry, 67*(3), 304–310.

Boyle, P. A., Buchman, A. S., Wilson, R. S., Yu, L., Schneider, J. A., & Bennett, D. A. (2012). Effect of purpose in life on the relation between Alzheimer disease pathologic changes on cognitive function in advanced age. *Archives of General Psychiatry, 69*(5), 499–504.

Brewer, M. B., & Gardner, W. (1996). Who is the "we"? Levels of collective identity and self representations. *Journal of Personality and Social Psychology, 71*(1), 83–93.

Bronk, K. C. (2012). The role of purpose in life in healthy identity formation: A grounded model. *New Directions for Youth Development, 132*, 31–44.

Bronk, K. C. (2013). *Purpose in life: A critical component of optimal youth development.* New York: Springer.

Bronk, K. C., Hill, P. L., Lapsley, D. K., Talib, T. L., & Finch, H. (2009). Purpose, hope, and life satisfaction in three age groups. *The Journal of Positive Psychology, 4*(6), 500–510. https://doi.org/10.1080/17439760903271439

Burrow, A. L., & Hill, P. L. (2011). Purpose as a form of identity capital for positive youth adjustment. *Journal of Developmental Psychology, 47*(4), 1196.

Burrow, A. L., O'Dell, A. C., & Hill, P. L. (2010). Profiles of a developmental asset: Youth purpose as a context for hope and well-being. *Journal of Youth and Adolescence, 39*, 1265–1273. https://doi.org/10.1007/s10964-009-9481-1

Damon, W., Menon, J., & Bronk, K. C. (2003). The development of purpose during adolescence. *Applied Developmental Science, 7*(3), 119–128.

Fredrickson, B. L., Grewen, K. M., Algoe, S. B., Firestine, A. M., Arevalo, J. M., Ma, J., & Cole, S. W. (2015). Psychological well-being and the human conserved transcriptional response to adversity. *PLoS One, 10*(3), e0121839.

Friedman, E. M., Hayney, M., Love, G. D., Singer, B. H., & Ryff, C. D. (2007). Plasma interleukin-6 and soluble IL-6 receptors are associated with psychological well-being in aging women. *Health Psychology, 26*(3), 305.

Hertzog, C., Kramer, A. F., Wilson, R. S., & Lindenberger, U. (2008). Enrichment effects on adult cognitive development: Can the functional capacity of older adults be preserved and enhanced? *Psychological Science in the Public Interest, 9*(1), 1–65.

Hill, P. L., & Burrow, A. L. (2012). Viewing purpose through an Eriksonian lens. *Journal of Identity, 12*(1), 74–91.

Hill, P. L., Burrow, A. L., Brandenberger, J. W., Lapsley, D. K., & Quaranto, J. C. (2010). Collegiate purpose orientations and well-being in early and middle adulthood. *Journal of Applied Developmental Psychology, 31*(2), 173–179.

Hill, P. L., Burrow, A. L., & Bronk, K. C. (2016). Persevering with positivity and purpose: An examination of purpose commitment and positive affect as predictors of grit. *Journal of Happiness Studies, 17*(1), 257–269. https://doi.org/10.1007/s10902-014-9593-5

Hill, P. L., Burrow, A. L., O'Dell, A. C., & Thornton, M. A. (2010). Classifying 'adolescents' conceptions of purpose in life. *The Journal of Positive Psychology, 5*(6), 466–473.

Hill, P. L., Edmonds, G. W., & Hampson, S. E. (2017). A purposeful lifestyle is a healthful lifestyle: Linking sense of purpose to self-rated health through multiple health behaviors. *Journal of Health Psychology*, 1359105317708251.

Hill, P. L., Edmonds, G. W., Peterson, M., Luyckx, K., & Andrews, J. A. (2016). Purpose in life in emerging adulthood: Development and validation of a new brief measure. *The Journal of Positive Psychology, 11*(3), 237–245.

Hill, P. L., Sumner, R., & Burrow, A. L. (2014). Understanding the pathways to purpose: Examining personality and well-being correlates across adulthood. *The Journal of Positive Psychology, 9*(3), 227–234.

Hill, P. L., & Turiano, N. A. (2014). Purpose in life as a predictor of mortality across adulthood. *Psychological Science, 25*(7), 1482–1486.

Hill, P. L., Turiano, N. A., Mroczek, D. K., & Burrow, A. L. (2016). The value of a purposeful life: Sense of purpose predicts greater income and net worth. *Journal of Research in Personality, 65*, 38–42. https://doi.org/10.1016/j.jrp.2016.07.003

Hill, P. L., & Weston, S. J. (2019). Evaluating eight-year trajectories for sense of purpose in the health and retirement study. *Aging & Mental Health, 23*(2), 233–237.

Hofer, S. M., & Sliwinski, M. J. (2001). Understanding ageing. *Gerontology, 47*(6), 341–352.

Holahan, C. K., Holahan, C. J., & Suzuki, R. (2008). Purposiveness, physical activity, and perceived health in cardiac patients. *Disability and Rehabilitation, 30*(23), 1772–1778.

Hooker, S. A., & Masters, K. S. (2016). Purpose in life is associated with physical activity measured by accelerometer. *Journal of Health Psychology, 21*(6), 962–971.

Ishida, R., & Okada, M. (2006). Effects of a firm purpose in life on anxiety and sympathetic nervous activity caused by emotional stress: Assessment by psycho-physiological method. *Stress and Health: Journal of the International Society for the Investigation of Stress, 22*(4), 275–281.

Jung, T., & Wickrama, K. A. S. (2008). An introduction to latent class growth analysis and growth mixture modeling. *Social and Personality Psychology Compass, 2*(1), 302–317.

Karasawa, M., Curhan, K. B., Markus, H. R., Kitayama, S. S., Love, G. D., Radler, B. T., & Ryff, C. D. (2011). Cultural perspectives on aging and well-being: A comparison of Japan and the United States. *The International Journal of Aging and Human Development, 73*(1), 73–98. https://doi.org/10.2190/AG.73.1.d

Kashdan, T. B., & McKnight, P. E. (2009). Origins of purpose in life: Refining our understanding of a life well lived. *Psychological Topics, 18*, 303–316.

Kim, E. S., Hershner, S. D., & Strecher, V. J. (2015). Purpose in life and incidence of sleep disturbances. *Journal of Behavioral Medicine, 38*(3), 590–597.

Kim, E. S., Kawachi, I., Chen, Y., & Kubzansky, L. D. (2017). Association between purpose in life and objective measures of physical function in older adults. *JAMA Psychiatry, 74*(10), 1039–1045.

Kim, E. S., Strecher, V. J., & Ryff, C. D. (2014). Purpose in life and use of preventive health care services. *Proceedings of the National Academy of Sciences, 111*(46), 16331–16336.

Kim, E. S., Sun, J. K., Park, N., Kubzansky, L. D., & Peterson, C. (2013). Purpose in life and reduced risk of myocardial infarction among older US adults with coronary heart disease: A two-year follow-up. *Journal of Behavioral Medicine, 36*(2), 124–133.

Kim, E. S., Sun, J. K., Park, N., & Peterson, C. (2013). Purpose in life and reduced incidence of stroke in older adults: 'The health and retirement study'. *Journal of Psychosomatic Research, 74*(5), 427–432.

Ko, H. J., Hooker, K., Geldhof, G. J., & McAdams, D. P. (2016). Longitudinal purpose in life trajectories: Examining predictors in late midlife. *Psychology and Aging, 31*(7), 693–698.

Kroger, J., & Marcia, J. E. (2001). The identity statuses: Origins, meanings, and interpretations. In S. J. Schwartz, K. Luyckx, & V. Vignoles (Eds.), *Handbook of identity theory and research* (pp. 31–53). New York, NY: Springer.

Lewis, N. A., Brazeau, H., & Hill, P. L (2018). Adjusting after stroke: Changes in sense of purpose in life and the role of social support, relationship strain, and time. *Journal of Health Psychology*. https://doi.org/10.1177/1359105318772656

Lewis, N. A., & Hill, P. L. (2019). *Leisure activity engagement moderates the effect of retirement on decline in sense of purpose in life*. In *Manuscript in preparation*.

Lewis, N. A., Turiano, N. A., Payne, B. R., & Hill, P. L. (2017). Purpose in life and cognitive functioning in adulthood. *Aging, Neuropsychology, and Cognition, 24*(6), 662–671.

MacDonald, S. W., Hundza, S., Love, J. A., DeCarlo, C. A., Halliday, D. W., Brewster, P. W., … Dixon, R. A. (2017). Concurrent indicators of gait velocity and variability are associated with 25-year cognitive change: A retrospective longitudinal investigation. *Frontiers in Aging Neuroscience, 9*, 17. https://doi.org/10.3389/fnagi.2017.00017

Malin, H., Liauw, I., & Damon, W. (2017). Purpose and character development in early adolescence. *Journal of Youth and Adolescence, 46*(6), 1200–1215.

Mariano, J. M., & Savage, J. (2009). Exploring the language of youth purpose: References to positives states and coping styles by adolescents with different kinds of purpose. *Journal of Character Education, 7*(1), 1–24.

McKnight, P. E., & Kashdan, T. B. (2009). Purpose in life as a system that creates and sustains health and well-being: An integrative, testable theory. *Review of General Psychology, 13*(3), 242–251. https://doi.org/10.1037/a0017152

Pfund, G. N., & Hill, P. L. (2018). The multifaceted benefits of purpose in life. *The International Forum for Logotherapy, 41*, 27–37.

Pinquart, M. (2002). Creating and maintaining purpose in life in old age: A meta-analysis. *Ageing International, 27*(2), 90–114.

Ryff, C. D. (1989). Happiness is everything, or is it? Explorations on the meaning of psychological well-being. *Journal of Personality and Social Psychology, 57*, 1069–1081.

Ryff, C. D., Love, G. D., Urry, H. L., Muller, D., Rosenkranz, M. A., Friedman, E. M., … Singer, B. (2006). Psychological well-being and ill-being: Do they have distinct or mirrored biological correlates? *Psychotherapy and Psychosomatics, 75*(2), 85–95.

Schaefer, S. M., Boylan, J. M., Van Reekum, C. M., Lapate, R. C., Norris, C. J., Ryff, C. D., & Davidson, R. J. (2013). Purpose in life predicts better emotional recovery from negative stimuli. *PLoS One, 8*(11), e80329.

Sim, J. J., Goyle, E., McKedy, W., Eidelman, S., & Correll, J. (2014). How social identity shapes the working self-concept. *Journal of Experimental Social Psychology, 55*, 271–277.

Snyder, C. R., Rand, K. L., & Sigmon, D. R. (2005). Hope theory: A member of the positive psychology family. In C. R. Snyder & S. J. Lopez (Eds.), *Handbook of positive psychology* (pp. 257–276). New York, NY: Oxford University Press.

Springer, K. W., Pudrovska, T., & Hauser, R. M. (2011). Does psychological well-being change with age? Longitudinal tests of age variations and further exploration of the multidimensionality of 'Ryff's model of psychological well-being. *Social Science Research, 40*(1), 392–398.

Stern, Y. (2002). What is cognitive reserve? Theory and research application of the reserve concept. *Journal of the International Neuropsychological Society, 8*(3), 448–460.

Sumner, R., Burrow, A. L., & Hill, P. L. (2015). Identity and purpose as predictors of subjective well-being in emerging adulthood. *Emerging Adulthood, 3*(1), 46–54. https://doi.org/10.1177/2167696814532796

Wang, M. C., Lightsey, O. R., Pietruszka, T., Uruk, A. C., & Wells, A. G. (2007). Purpose in life and reasons for living as mediators of the relationship between stress, coping, and suicidal behavior. *The Journal of Positive Psychology, 2*(3), 195–204.

Weston, S. J., Lewis, N. A., & Hill, P. L. (2019). *Changes in sense of purpose and social support during older adulthood*. In *Manuscript submitted for publication*.

Windsor, T. D., Curtis, R. G., & Luszcz, M. A. (2015). Sense of purpose as a psychological resource for aging well. *Developmental Psychology, 51*(7), 975.

Zilioli, S., Slatcher, R. B., Ong, A. D., & Gruenewald, T. L. (2015). Purpose in life predicts allostatic load ten years later. *Journal of Psychosomatic Research, 79*(5), 451–457.

Chapter 4
Personality Disorders and Disordered Aging: Personality Pathology as Risk Factor for Unhealthy Aging

Patrick J. Cruitt

4.1 Introduction

Healthy aging is far more than the absence of pathology in later life. However, any adequate conceptualization of health requires attending to the pathological. In the cognitive domain, studying patients with dementia has led to breakthroughs in understanding the processes involved in adaptive cognitive aging. However, the field of personality and aging has yet to routinely study the maladaptive extremes of personality, instead focusing on normal-range traits framed in generally positive terms. This disconnect goes both ways, as personality pathology in later life has long been considered an understudied area in clinical psychology due to a variety of assessment and conceptual issues, and only recently has been receiving the attention it deserves (Cruitt & Oltmanns, 2018a; Van Alphen et al., 2015). This increased interest has been complicated by an ongoing debate within the personality pathology field about the best approach to defining personality disorder (PD) broadly speaking. The purpose of the current chapter is to provide a brief overview of the state of the literature on the definition, associated outcomes, and treatment of personality pathology in later life. First, I will discuss the issues involved in conceptualizing and diagnosing personality pathology across the lifespan. Then, I will examine the literature on the relationship between personality pathology and healthy aging outcomes, including mental and physical health, cognitive aging, and social functioning. Finally, I will explore potential mechanisms of recovery from

A grant from the National Institutes of Health (5 T32 AG000030-39) supported this work.

P. J. Cruitt (✉)
Department of Psychological and Brain Sciences, Washington University in St. Louis, St. Louis, MO, USA
e-mail: pcruitt@wustl.edu

P. L. Hill, M. Allemand (eds.), *Personality and Healthy Aging in Adulthood*, International Perspectives on Aging 26,
https://doi.org/10.1007/978-3-030-32053-9_4

personality pathology and treatment interventions in later life. By examining the ways by which personality may increase risk for *un*healthy aging, we might begin to better understand the mechanisms that lead to healthy aging outcomes. In addition, we may be able to identify ways to help those suffering from personality pathology to age well and live healthier in later life.

4.2 Defining Personality Pathology

The primary barriers to drawing out personality pathology's implications for healthy aging to date have been definitional in nature. Defining personality pathology is a difficult task, and one that the field has been grappling with for decades. The long-standing categorical model of diagnosis included in the main text (Section II) of the *Diagnostic and Statistical Manual of Mental Disorders, Fifth Edition* (*DSM-5*) has a number of limitations that interfere with studying personality pathology and healthy aging. These limitations include both general critiques of the model that apply across the lifespan, as well as specific issues regarding the conceptualization of PDs in later life. In order to address some of these limitations, diagnostic models based around dimensional assessments of personality traits have been proposed as both the Alternative Model for PDs (AMPD) in Section III of the *DSM-5* and the upcoming 11th revision to the *International Statistical Classification of Diseases and Related Health Problems* (*ICD-11*; American Psychiatric Association, 2013; World Health Organization, 2018). Despite the fact that the field is in a state of flux, some general PD criteria remain consistent across all three models. A PD is defined as a maladaptive personality pattern that is (A) relatively stable, (B) inflexible, (C) pervasive across situations, (D) not culturally or developmentally appropriate, and (E) associated with significant distress and/or impairment in functioning. Where the models differ is the way in which they characterize the type and severity of the personality pattern in question, using either discrete categories or dimensions. The transition to dimensional approaches has the potential to enhance our understanding of the relationship between personality pathology and healthy aging through extensions of the adaptive-range personality trait literature and providing more appropriate assessments of how the maladaptive extremes of personality manifest in older adulthood.

Much of the research on PDs has used (and continues to use) the Section II categorical model, even as researchers have advocated for transitioning to dimensions for decades. This model consists of ten categorical diagnoses, separated into three clusters on the basis of similar features. Cluster A (paranoid, schizoid, and schizotypal PD) is characterized by odd/eccentric features, Cluster B (antisocial, borderline, histrionic, and narcissistic PD) by dramatic/erratic features, and Cluster C (avoidant, dependent, and obsessive-compulsive PD) by anxious/fearful features. Each specific PD requires an individual to meet a certain threshold of the relevant criteria for diagnosis. Decades of research on this model have identified serious limitations, with important implications for the study of healthy aging. Extensive heterogeneity within and comorbidity among the PD diagnoses (Zimmerman, Rothschild, & Chelminski, 2005) and the over-use of personality disorder not otherwise specified as a diagnosis in clinical settings (Verheul, Bartak, & Widiger, 2007; Verheul & Widiger, 2004) has led researchers to conclude that these categories fail to capture discrete syndromes that represent the full range of personality pathology. Also problematic is that the categories demonstrate relative *instability*, comparable to other forms of psychopathology and contrary to the general definition of PD (Clark, 2007; McDavid & Pilkonis, 1996). That is, longitudinal studies of PDs consistently show that the vast majority of patients with PDs will frequently experience a remission of symptoms such that they no longer meet criteria for the disorder (Keuroghlian & Zanarini, 2015). However, despite changes in diagnostic status, these patients continue to exhibit impairment in important life domains, such as work (Keuroghlian & Zanarini, 2015; Skodol et al., 2005). These findings illustrate the fact that the thresholds for diagnoses are arbitrary, and additional evidence suggests that even subthreshold symptoms show associations with distress/impairment (Ellison, Rosenstein, Chelminski, Dalrymple, & Zimmerman, 2016). As individuals enter later life, excessive comorbidity, diagnostic instability, and the information lost due to adhering to arbitrary thresholds make disentangling the effects of personality pathology on healthy aging outcomes increasingly difficult.

These general issues with the categorical model are only compounded by problems that are specific to the context of later life. First, some of the criteria for Section II PD diagnoses hold little face validity for older adults (Oltmanns & Balsis, 2011; Van Alphen et al., 2015). An oft-used example is the avoidant PD criterion "avoids occupational activities," which makes little sense for someone who is retired. Item-response theory analyses that examined age bias in the PD criteria identified a number of criteria that older adults (aged 65+) were more or less likely to endorse than younger adults with the same level of pathology (Balsis, Gleason, Woods, & Oltmanns, 2007). Second, the requirement that the PD criteria be present since young adulthood makes assessment difficult and precludes the diagnosis of personality pathology that may develop in later life. This requirement persists despite agreement among experts that "late onset" PD is possible, particularly in the context of poor adaptation to a later life transition such as the death of a spouse (Rosowsky, Lodish, Ellison, & Van Alphen, 2019). Although the concept of late onset PD needs to be empirically validated, such research requires removing the a priori requirement found in *DSM-5*. Taken alongside the issues with the categorical Section II model described above, the age-specific problems present a twofold challenge for the gerontologist interested in personality pathology.

Fortunately, both the AMPD and the ICD-11 model seek to address the limitations of the categorical model by proposing dimensional models based on the normal-range personality trait literature. Dimensional models exhibit greater stability and are able to capture and preserve a wider range of information than categorical diagnoses. Both models also separate out an assessment of severity from the particular stylistic manifestation of an individual's personality pathology, helping to address concerns about arbitrary diagnostic thresholds and comorbidity (Krueger, Hopwood, Wright, & Markon, 2014; Tyrer et al., 2011). In both models, a clinician rates the overall severity of impairment in personality functioning, using either a scale from mild to severe (ICD-11) or one that includes specific descriptions of impairment in both self- and interpersonal functioning (the Level of Personality Functioning Scale in the AMPD). Then, individuals receive a trait profile using five domains that for the most part represent maladaptive extensions of the normal-range Big Five. For example, the domain of negative affectivity represents the pathologically high end of neuroticism, and is characterized by features such as emotional lability, depressivity, and distrust. Three other domains are similar across the AMPD and ICD-11: Detachment (low extraversion), antagonism/dissociality (low agreeableness), and disinhibition (low conscientiousness). The two models differ in their fifth domain: the AMPD includes psychoticism to capture oddity/eccentricity (somewhat related to high openness), whereas ICD-11includes an anankastia domain to capture rigidity/perfectionism (high conscientiousness). Overall, research shows that the AMPD domains line up empirically with the Big Five as expected (Gore & Widiger, 2013; Krueger & Markon, 2014), and that measures of these constructs capture additional information at the extremes of the same underlying dimensions as the Big Five (Suzuki, Samuel, Pahlen, & Krueger, 2015). As such, the new dimensional models represent powerful tools to expand upon research on the relationship between the Big Five and healthy aging and begin to identify mechanisms that may link the extremes of personality to particular aging outcomes.

4.2.1 Dimensional Models in Older Adulthood

However, even though addressing the limitations of the categorical approach was a major goal of the development of the AMPD and ICD-11 models, it is unclear to what extent attempts were made to address age-related concerns. Research on the dimensional models in later life remains surprisingly scant. There is evidence that the structure of the AMPD personality trait model may be replicated in later life (Van Den Broeck et al., 2014). With regard to addressing the age-bias concerns with the categorical model, dimensional models have inherent advantages, as age bias in any individual item is not likely to dramatically alter diagnosis. Even measures of the Section II PD symptoms exhibit less age bias when scored dimensionally (Debast et al., 2015). Nevertheless, there is evidence that certain lower-order traits of the AMPD model, particularly withdrawal, attention seeking, rigid perfectionism and unusual beliefs, exhibit age bias (Van Den Broeck, Bastiaansen, Rossi, Dierckx, & De Clercq, 2013). Unfortunately, the AMPD trait model also continues to include the requirement that the maladaptive personality trait profile be present since young adulthood. However, the ICD-11 model does not and is therefore consistent with the concept of late onset PD (Tyrer et al., 2011; Videler, Hutsebaut, Schulkens, Sobczak, & Van Alphen, 2019). Further research is necessary to examine how the dimensional models perform in the context of later life, and the clinical utility of using these diagnostic systems with older adults. Doing so will help to refine the definition of personality pathology and place it within the context of a lifespan perspective.

4.2.2 *Epidemiological Differences Between Young and Older Adulthood*

Once personality pathology has been defined, the natural next question is how prevalent it is in older adulthood. Almost all of the epidemiological work to date has used the Section II categorical model. Clinical wisdom and early empirical evidence suggested that rates of PDs declined with age (Cohen et al., 1994). However, studies in general show similar prevalence rates of PDs in later life (at least in individuals age 50 and older) as in younger adulthood, with approximately ten percent of the population meeting criteria for a PD diagnosis (Abrams & Horowitz, 1996, 1999; Oltmanns, Rodrigues, Weinstein, & Gleason, 2014). There is some evidence from the National Epidemiologic Survey on Alcohol and Related Conditions that prevalence rates after age 65 are similar (Schuster, Hoertel, Le Strat, Manetti, & Limosin, 2013), although these rates may be based on less stringent diagnostic criteria (Trull, Jahng, Tomko, Wood, & Sher, 2010). As PD instruments developed with the later life context in mind see more widespread implementation, it will be important to obtain more detailed epidemiological estimates. It will also be important to obtain estimates using dimensional models. Regardless, the current findings suggest that personality pathology continues to hold relevance in later life.

Although the overall prevalence rate of any PD diagnosis appears similar in both younger and older adulthood, the rates for specific diagnoses are more variable. As mentioned above, paranoid, schizoid and obsessive-compulsive symptoms may be more common in later life (Abrams & Horowitz, 1999), whereas certain borderline and antisocial symptoms may be less common (Samuels et al., 2002). Even within a particular disorder, symptom presentations may change with age. There is evidence that personality pathology demonstrates heterotypic continuity, such that behavioral manifestations associated with similar levels of the latent maladaptive traits differ across the lifespan. This idea has been most extensively studied with regard to borderline PD, with studies demonstrating that impulsivity and self-harm behaviors decrease with age whereas chronic emptiness and somatic complaints increase (see Beatson et al., 2016, for a review). These findings suggest that the mechanisms linking PDs with health outcomes may differ across the lifespan, and that it is particularly important to conduct careful assessments of PD features in later life. Overall, the evidence suggests that personality pathology, though somewhat changed in manifestation, persists into later life.

4.3 Personality Pathology and Healthy Aging Outcomes

Given that personality pathology remains a relevant issue in older adulthood, understanding the ways in which maladaptive personality patterns represent risk factors for unhealthy aging outcomes is essential to both prediction and intervention. Evidence suggests that the maladaptive extremes of personality contribute additional variance above and beyond normal-range personality traits to the prediction of social functioning, physical health, and (lack of) stressful life events in later life (Gleason, Weinstein, Balsis, & Oltmanns, 2014). This incrementation may not only aid prediction, but also help illuminate the mechanisms by which personality impacts healthy aging and identify targets for intervention. Behaviors associated with the maladaptive extreme ends of underlying personality trait dimensions may yield insight into the relatively modest effects of normal-range personality traits on various healthy aging outcomes, such as cognitive decline. The following sections review the existing literature on the relationship between personality pathology and relevant healthy aging outcomes in the domains of mental health, physical health, cognitive aging, and interpersonal functioning. However, first I will explore the relationship between personality pathology and two multiply determined lifespan outcomes with particular relevance for healthy aging: mortality and work disability.

4.3.1 All-Cause Mortality, Work Disability and Personality Pathology

The presence of personality pathology across the lifespan has many important implications, although perhaps none so important as predicting who survives into older adulthood in the first place. All-cause mortality is higher for patients with PDs compared to the general population, and PDs are associated with some of the highest risks among mental health conditions for predicting mortality due to medical conditions and suicide (Nordentoft et al., 2013). This increased mortality risk is associated with a decrease in life expectancy relative to the general population, with gender-stratified estimates ranging from 13 to 19 years lower on average (Fok et al., 2012; Nordentoft et al., 2013). Although life expectancy is an important outcome in its own right, living longer does not always mean that one is doing so with a high quality of life. Researchers must also examine the role personality pathology plays in lifespan outcomes with implications for the quality of life in older adulthood, such as work disability.

Besides the clear implications of lost income for later-life outcomes, research shows that older adults experience greater declines in life satisfaction and poorer adaptation to work disability relative to middle adults (Infurna & Wiest, 2018). PDs are associated with greater risk for being placed on disability pensioning earlier in life (Amundsen Østby et al., 2014; Knudsen et al., 2012). Individuals with borderline PD in particular show high rates of receiving social security disability income benefits over time (Zanarini, Jacoby, Frankenburg, Reich, & Fitzmaurice, 2009). As individuals enter retirement age, work-related disability may manifest as early retirement. PDs and depression are both associated with a comparably high risk for early retirement on health grounds, more so than anxiety disorders (Korkeila et al., 2010). Previous research suggests that involuntary retirement is associated with poor physical and mental health outcomes (Hershey & Henkens, 2014; Hyde,

Hanson Magnusson, Chungkham, Leineweber, & Westerlund, 2015; Mosca & Barrett, 2016; van Solinge, 2007). Therefore, early exit from the work force may be an important outcome in later life just as it is in young and middle adulthood.

4.3.2 Mental Health and Personality Pathology

There may be a number of different pathways linking personality pathology to all-cause mortality and work disability, through a variety of mental and physical health comorbidities. Fortunately, a large literature examines the relationship between personality pathology and other mental and physical health conditions. PDs exhibit extensive comorbidity with other psychiatric disorders across the lifespan (P. Cohen, Crawford, Johnson, & Kasen, 2005; Skodol et al., 1999). Depression in later life is particularly devastating, given that the typical mean-level trajectory associated with aging is one of increasing well-being and life satisfaction (see Katana & Hill, this volume). Not only is personality pathology such as borderline PD associated with a history of major depression in older adults (Galione & Oltmanns, 2013), it substantially interferes with recovery from depression. In small samples of patients who no longer met criteria for depression after antidepressant treatment, PD symptoms interacted with remaining depressive symptoms to predict worse functioning at low levels of residual depression, both cross-sectionally and over a year of follow-up (Abrams, Alexopoulos, Spielman, Klausner, & Kakuma, 2001; Abrams, Spielman, Alexopoulos, & Klausner, 1998). Cluster C PDs have been shown to interfere with treatment, predicting longer time-to-response in acute treatment and non-response in continuation and maintenance phases (Morse, Pilkonis, Houck, Frank, & Reynolds, 2005). These findings suggest that older patients presenting with depression should be screened for personality pathology. As will be discussed below, researchers have begun testing the effectiveness of interventions originally designed to target PDs for treating older adults with comorbid depression. Adapting PD-focused treatments for use in this population may help aid in recovery from depression and promote healthy aging.

Personality pathology may also be an important target for preventative interventions for suicide. Suicidal behavior is a complex mental health phenomenon, and one that takes on an added significance in later life. Rates of death by suicide increase in older adulthood, particularly for men (Nock et al., 2008; Shah, Bhat, Zarate-Escudero, Deleo, & Erlangsen, 2016). There is evidence indicating that borderline PD features and neuroticism are uniquely linked to increased suicidal ideation in older adults (Segal, Marty, Meyer, & Coolidge, 2012). One study suggests that measures of depressive symptoms, thwarted belongingness and perceived burdensomeness account for some of the variance in the relationship between PD features and suicidal ideation (Jahn, Poindexter, & Cukrowicz, 2015). Although the cross-sectional nature of these findings precludes any conclusions about causality, these interpersonal functioning outcomes may be one mechanism linking PD features with suicidal behaviors. As such, aging researchers would benefit from including measures of belongingness, burdensomeness, and other markers of interpersonal functioning in prospective studies of this relationship. Beyond suicidal ideation, PD features are linked to risk for dying by suicide in later life (Harwood, Hawton, Hope, & Jacoby, 2001). As with mortality due to other causes, personality pathology appears to play an important role in predicting who lives into older adulthood. Interventions to address personality pathology, either directly or through their influence on interpersonal risk factors, may not just prolong life through suicide prevention but also improve the quality of life that individuals experience while aging.

4.3.3 Physical Health and Personality Pathology

In addition to mental health outcomes, personality pathology shows negative associations with a wide variety of physical health outcomes, including several age-related conditions (Dixon-Gordon, Whalen, Layden, & Chapman, 2015; Quirk et al., 2016). For instance, analyses of data from National Epidemiologic Survey on Alcohol and Related Conditions find cross-sectional relationships between many of the Section II PDs and arthritis, as well as between antisocial and borderline PD and cardiovascular disease (El-Gabalawy, Katz, & Sareen, 2010; Goldstein et al., 2008; McWilliams, Clara, Murphy, Cox, & Sareen, 2008). A longitudinal study confirmed that Cluster B PDs (e.g., borderline and antisocial PD) predict future incidence of cardiovascular disease (Lee et al., 2010). These relationships take on added significance in later life, as chronic medical conditions become more prevalent and the burden on the health care system is magnified (Newton-Howes, Clark, & Chanen, 2015; Oltmanns & Balsis, 2011). Total PD features predict decreased physical functioning and increased medication and health care usage after 6 months of follow-up in late middle adulthood (Powers & Oltmanns, 2012), and antisocial and narcissistic PD features uniquely predict greater healthcare utilization over the course of 2 years when controlling for number of physical health problems (Powers, Strube, & Oltmanns, 2014).

Given these findings, it is important to identify the mechanisms by which maladaptive personality traits impact physical health outcomes in older adulthood to promote healthy aging. For example, obesity appears to account for much of the variance in the relationship between BPD, in particular the impulsivity component of BPD, and physical health conditions like arthritis and heart disease (Iacovino, Powers, & Oltmanns, 2014; Powers & Oltmanns, 2013). Although these findings need to be replicated longitudinally in order to determine the direction of causality, they suggest that weight loss interventions in mid- to late life may be helpful in mitigating the negative effect of borderline PD on physical health, if they can be appropriately tailored to address the patient's level of impulsivity. Future research is needed to examine additional possible mediators, such as exercise or other health behaviors.

4.3.4 Cognitive Aging and Personality Pathology

Although mental and physical health are important components of healthy aging, changes in cognitive functioning are perhaps the most prototypical outcomes of the aging process. Given the extensive literature on the association between normal-range personality traits and cognitive aging, it is surprising that personality pathology has rarely been studied with regard to cognitive aging outcomes. Fortunately, the normal-range personality literature can provide insight into the role personality pathology may play in cognitive aging. Previous studies have demonstrated that personality traits associated with PDs, namely high neuroticism and low conscientiousness, prospectively predicts declines in certain cognitive abilities (Curtis, Windsor, & Soubelet, 2015; Luchetti, Terracciano, Stephan, & Sutin, 2016) and risk for mild cognitive impairment or dementia (Low, Harrison, & Lackersteen, 2013; Terracciano et al., 2014; Terracciano, Stephan, Luchetti, Albanese, & Sutin, 2017). However, these associations between normal-range personality and cognitive aging have been relatively modest. As mentioned above, examining the maladaptive extremes of personality may enhance the prediction of this important outcome, and illuminate potential mechanisms.

A few studies have examined the association between PD features and symptoms of dementia or cognitive decline. These studies have highlighted an important issue: There is substantial overlap between PD criteria and non-cognitive symptoms of dementia. For example, one study of patients with a diagnosis of probable Alzheimer's disease (AD) found associations between retrospective informant reports of PD symptoms and non-cognitive dementia symptoms (Prior et al., 2016). As one might expect, PD features were associated with dementia symptoms that overlapped with the PD criteria themselves. Premorbid Cluster A (odd/eccentric) features, which include intense social anxiety/withdrawal and unusual perceptual experiences, were associated with anxiety, depression, and hallucinations. Cluster B (dramatic/erratic) features showed associations with aggression and irritability, which are also part of the criteria of Cluster B PDs. Finally, Cluster C (anxious/fearful) features, including feelings of inadequacy and dependency, were associated with depression. Another study examined a specific Cluster C PD, dependent PD, and its association with a diagnosis of mild cognitive impairment or dementia (Pilleron et al., 2015). However, the dependent PD criteria may not be valid indicators of pathology in the context of cognitive decline, during which increased dependency is normative and potentially adaptive. These findings suggest that differentiating between dementia and PD in later life may be difficult in the early stages of the disease, before declines in cognitive ability become apparent. Case studies support this conclusion, illustrating the difficulty of differential diagnosis in clinical settings (Greve, Curtis, & Bianchini, 2007; Greve, Curtis, Bianchini, & Collins, 2004; Hellwig, Dykierek, Hellwig, Zwernemann, & Meyer, 2012; Helmes & Steward, 2010). As such, researchers and clinicians must carefully assess symptoms to establish whether they represent long-standing aspects of the patient's personality, or whether they have appeared *de novo* as a consequence of disease. It is possible that personality pathology earlier in life influences the manifestation of behavioral symptoms associated with dementia. At the same time, it is also possible that dementia gives rise to new, pathological manifestations of personality traits that were well-regulated prior to disease onset. Future research will need to explore both of these possibilities.

Beyond issues of confounding symptoms, there does appear to be some relationship between PD features and cognitive decline. A study of patients with a probable AD diagnosis showed that informants' retrospective reports of the patient's premorbid personality indicated higher levels of PD features than unaffected siblings (Nicholas et al., 2010). This study suggests that PD features in middle adulthood predict future dementia, although it is possible that informants' retrospective reports were biased by the patient's current functioning. Another study examined obsessive-compulsive PD specifically and found an association with AD (Dondu, Sevincoka, Akyol, & Tataroglu, 2015). However, interviewers integrated information from both patient and informant report in determining if patients met criteria for obsessive-compulsive PD, out of fears that the patient report would be unreliable. If patient diagnoses of obsessive-compulsive PD were more dependent on information obtained from informants than control diagnoses were, this may lead to biased results as a consequence of informants reporting more or less pathology than the self (Oltmanns et al., 2014; Sleep, Lamkin, Lynam, Campbell, & Miller, 2018). One study has addressed this issue by testing the incremental validity of self-, informant, and interview reports of current personality pathology in predicting informant reported changes in cognition (Cruitt & Oltmanns, 2018b). PDs were meaningfully associated with changes in cognition, and the shared variance among the sources of information accounted for much of this relationship. In addition, each of the three sources of information contributed unique information to the prediction of cognitive change. These findings suggest that a multi-source approach to the assessment of PDs in later life may help clarify relationships with cognitive aging outcomes. Although progress has been made, it remains to be seen whether PDs longitudinally predict cognitive decline, and through what mechanisms.

4.3.5 Interpersonal Functioning and Personality Pathology

Interpersonal functioning in later life is perhaps the domain with the most potential for identifying mechanisms linking personality pathology to cognitive decline and other healthy aging outcomes. Social engagement and social support are consistently linked to healthy cognitive aging (Hertzog, Kramer, Wilson, & Lindenberger, 2009; Smith, 2016), physical health (S. Cohen, 2004), and mental health (Kawachi & Berkman, 2001). Although losses lead to smaller social network sizes in older adulthood, there is evidence that increased relationship satisfaction is an important, normative outcome of healthy aging (Luong, Charles, & Fingerman, 2011). This trajectory appears to be due to increased motivation to cultivate positive social interactions. There exist a number of reasons to expect personality pathology to interfere with this process. First, interpersonal dysfunction is part of the definition of PD. Many specific Section II PDs include criteria directly impinging on interpersonal relationships, such as borderline PD criteria about intense, unstable relationships and fear of abandonment. By the same token, both dimensional models include interpersonal dysfunction as a core part of the severity dimension of personality pathology. However, even though this definitional overlap exists, it is important to establish that personality pathology predicts particular relationship outcomes over time, and whether these relationship outcomes account for some of the relationship between PDs and other healthy aging outcomes. Research indicates that PD features, particularly borderline PD features, prospectively predict increased rates of dependent, interpersonal stressful life events in late middle adulthood (Conway, Boudreaux, & Oltmanns, 2018; Gleason, Powers, & Oltmanns, 2012; Powers, Gleason, & Oltmanns, 2013). This interpersonal stress generation likely has negative downstream consequences for health outcomes, as individuals potentially engage in fewer positive social activities and perceive less support from close others. As such, interpersonal functioning in the context of personality pathology and healthy aging deserves research attention not only as an important outcome in its own right, but also as a possible mediating factor.

4.4 Mechanisms for Recovery from Personality Pathology

All of the literature on personality pathology as a risk factor for unhealthy aging outcomes reviewed above leaves open a critical question: What does healthy aging look like for a person with maladaptive personality traits? From the perspective of the clinical literature, the first step toward addressing this question is to test the effectiveness of treatment interventions. Relatively few studies have examined the effectiveness of PD treatments specifically for older adult populations. However, researchers and clinical experts have identified some general principles for working with older patients who have a PD diagnosis (Van Alphen et al., 2012; Videler et al., 2015). In particular, they recommend multiple levels of care dependent on the patient's insight, motivation, and capacity to change. Highly motivated patients with good insight may receive treatment aimed at personality change, whereas treatment for those patients with poor insight or limited capacity to change may focus on adaptation to their current environment. Finally, treatment for patients who face significant barriers to change may focus on providing structure and support. Although these general guidelines could apply to a number of different types of psychotherapy, only two have been empirically tested for use with older adults suffering from personality pathology: dialectical behavior therapy and schema therapy.

Dialectical behavior therapy (DBT) is an empirically supported treatment for borderline PD that integrates cognitive-behavioral techniques with acceptance-based practices (Linehan, 1993). One of the core features of DBT that separates it from other treatment approaches is group-based skills training, focused on developing mindfulness, interpersonal effectiveness, emotion regulation, and distress tolerance skills. One study examined the effectiveness of DBT plus medication for treating older adults with both depression and PD features (Lynch et al., 2007). Patients who received DBT plus medication experienced a more rapid remission in depressive symptoms than those who received medication alone. Other studies have begun to examine the effectiveness of schema therapy (ST), a therapeutic approach aimed at identifying and modifying patterns of behavior derived from core beliefs (or schemas) about the self, others, and the world formed early in life (Khasho et al., 2019; Videler, Rossi, Schoevaars, Van Der Feltz-Cornelis, & Van Alphen, 2014). Therapists use a variety of cognitive and experiential strategies, as well as the therapeutic relationship itself, to modify maladaptive schemas and replace them with healthier alternatives. A pre/post design proof of concept study found that group-based ST appears to have a beneficial effect for older patients with PD features or longstanding mood disorder (Videler et al., 2014). Specifically, patients experienced a decrease in psychological symptoms, and that change in distress later in treatment was associated with earlier change in maladaptive schemas. Research on specialized treatments for PDs in older adulthood is in its infancy. However, there is evidence to suggest that existing treatments can be adapted to the particular context of later life and help decrease psychological distress. Future research is required to examine whether these treatment effects generalize to the healthy aging outcomes discussed above and promote genuine flourishing.

Whereas studies of treatment efficacy are the purview of clinical psychology, there are a number of ways that more basic aging research can expand our understanding of the role different contexts play in helping individuals with personality pathology to experience healthier aging outcomes. Studies of younger patients with borderline PD diagnoses over time have identified a number of potential mechanisms of recovery, such as a stable romantic relationship or employment (Paris, 2003). However, little is known about how the loss of these factors through normative later life transitions (e.g., the death of a partner or retirement) impacts functioning (Beatson et al., 2016; Hutsebaut et al., 2019). It may be that the maintenance of these factors in later life buffers against negative outcomes. For example, one study found that employment status moderated the cross-sectional relationship between borderline PD and self- and informant report of physical health (Cruitt, Boudreaux, Jackson, & Oltmanns, 2018). Further research is needed to determine how this moderating effect plays out over time, and whether retirement functions differently than unemployment in this regard.

4.5 Implications for Aging Research

Personality pathology and healthy aging research is at an exciting turning point. As the PD field resolves ongoing disputes about definitional issues, particularly age-related bias and excluding the possibility of late-onset PD, it will become easier to incorporate measures of personality pathology into studies on aging. That being said, the current state of the literature offers important insight into how people without personality pathology tend to experience healthier aging. On average, they live longer and are less likely to experience work disability over the course of the lifespan, leading to greater quality of life in older adulthood. Individuals with more adaptive personalities are at lower risk for other psychiatric disorders and show better response to treatment for major depressive disorder. They are less likely to experience suicidal ideation and die by suicide in late life. Risk for age-related medical conditions is lower for those without personality pathology, and it appears that this may be true for dementia as well. The next stage of research will need explore the mechanisms that account for these associations with healthy aging outcomes, such as personality pathology's impact on interpersonal stress and engagement. Finally, there appears to be hope for individuals suffering from personality pathology. Treatment approaches specifically for use with older adults are in the process of being developed and tested. Furthermore, certain contexts may promote recovery and buffer against negative outcomes, such as employment. These conclusions suggest that the field as a whole has started to move beyond personality pathology as a risk factor, and toward an understanding of ways to promote healthy aging in individuals with unhealthy personalities.

4.5.1 Future Directions

Nevertheless, there are numerous gaps in the literature to be filled by future research. The field of clinical psychology must continue exploring treatment effectiveness and identifying options for clinicians to use to address the various levels of PD treatment in later life. Currently, an ongoing multiple baseline study seeks to illustrate the effectiveness of individual ST in older individuals with either a categorical diagnosis of borderline PD or subthreshold symptoms of borderline PD alongside elevated AMPD trait scores (Khasho et al., 2019). This study stands to represent a significant step forward in that it attempts to translate across categorical and dimensional models. It is also one of the first studies to examine treatment effectiveness for personality disorders in older adults who are not suffering from comorbid mood disorder. Although the field stands to learn a lot from this and other studies, it will be necessary in the future to address whether treatment not only decreases psychological distress, but also produces better outcomes in other healthy aging domains. Doing so will help integrate findings across the clinical and aging literatures.

More basic aging research also has the opportunity to explore pathways to healthy aging for individuals suffering from personality pathology. For example, exploring the possible mediating role of interpersonal functioning may identify potential interventions at the level of relationships. It may be that preventing the interpersonal stress generation associated with PDs produces better mental, physical, and cognitive health outcomes. Another potential future direction for aging research is to expand upon the existing evidence suggesting that certain life contexts (e.g., employment, romantic relationships) play a moderating role on the relationship between personality pathology and healthy aging. In doing so, it will be important to explore the unfolding of these contexts over time. What happens when an individual with borderline PD whose job provided them with a stable context and sense of self loses their job or enters retirement? Conversely, are there new contexts that arise in later life that may also play a salutary role, like grandparenthood? Along a similar vein, aging research stands to discover non-clinical interventions that positively impact aging outcomes through their influence on personality pathology. Such research has enormous potential to help improve older adult's lives. Not only will it provide a sense of how to help those with PDs live healthier lives, but it will also uncover mechanisms leading to healthy aging broadly speaking. Through the study of pathology, a clearer understanding of health is obtained.

Acknowledgments I would like to thank Patrick Hill and Ellie Martin for their helpful feedback on this manuscript.

References

Abrams, R. C., Alexopoulos, G. S., Spielman, L. A., Klausner, E., & Kakuma, T. (2001). Personality disorder symptoms predict declines in global functioning and quality of life in elderly depressed patients. *American Journal of Geriatric Psychiatry, 9*(1), 67–71. https://doi.org/10.1097/00019442-200102000-00010

Abrams, R. C., & Horowitz, S. V. (1996). Personality disorders after age 50: A meta-analysis. *Journal of Personality Disorders, 10*(3), 271–281. https://doi.org/10.1521/pedi.1996.10.3.271

Abrams, R. C., & Horowitz, S. V. (1999). Personality disorders after age 50: A meta-analytic review of the literature. In E. Rosowsky, R. C. Abrams, & R. A. Zweig (Eds.), *Personality disoders in older adults: Emerging issues in diagnosis and treatment* (pp. 55–68). Mahwah, NJ: Lawrence Erlbaum Associates Publishers.

Abrams, R. C., Spielman, L. A., Alexopoulos, G. S., & Klausner, E. (1998). Personality disorder symptoms and functioning in elderly depressed patients. *The American Journal of Geriatric Psychiatry, 27*(10), 1035–1045. https://doi.org/10.1097/00019442-199802000-00004

American Psychiatric Association. (2013). *Diagnostic and statistical manual of mental disorders* (5th ed.). Washington, DC: Author.

Amundsen Østby, K., Czajkowski, N., Knudsen, G. P., Ystrom, E., Gjerde, L. C., Kendler, K. S., … Reichborn-Kjennerud, T. (2014). Personality disorders are important risk factors for disability pensioning. *Social Psychiatry and Psychiatric Epidemiology, 49*(12), 2003–2011. https://doi.org/10.1007/s00127-014-0878-0

Balsis, S., Gleason, M. E. J., Woods, C. M., & Oltmanns, T. F. (2007). An item response theory analysis of DSM-IV personality disorder criteria across younger and older age groups. *Psychology and Aging, 22*(1), 171–185. https://doi.org/10.1037/0882-7974.22.1.171

Beatson, J., Broadbear, J. H., Sivakumaran, H., George, K., Kotler, E., Moss, F., & Rao, S. (2016). Missed diagnosis: The emerging crisis of borderline personality disorder in older people. *The Australian and New Zealand Journal of Psychiatry, 50*(12), 1139–1145. https://doi.org/10.1177/0004867416640100

Clark, L. A. (2007). Assessment and diagnosis of personality disorder: Perennial issues and an emerging reconceptualization. *Annual Review of Psychology, 58*(1), 227–257. https://doi.org/10.1146/annurev.psych.57.102904.190200

Cohen, B. J., Nestadt, G., Samuels, J. F., Romanoski, A. J., McHugh, P. R., & Rabins, P. V. (1994). Personality disorder in later life: A community study. *British Journal of Psychiatry, 165*(OCT), 493–499. https://doi.org/10.1192/bjp.165.4.493

Cohen, P., Crawford, T. N., Johnson, J. G., & Kasen, S. (2005). The children in the community study of developmental course of personality disorder. *Journal of Personality Disorders, 19*(5), 466–486. https://doi.org/10.1521/pedi.2005.19.5.466

Cohen, S. (2004). Social relationships and health. *American Psychologist, 59*(8), 676–684. https://doi.org/10.1037/0003-066X.59.8.676

Conway, C. C., Boudreaux, M., & Oltmanns, T. F. (2018). Dynamic associations between borderline personality disorder and stressful life events over five years in older adults. *Personality Disorders, Theory, Research, and Treatment, 9*(6), 521–529. https://doi.org/10.1037/per0000281

Cruitt, P. J., Boudreaux, M. J., Jackson, J. J., & Oltmanns, T. F. (2018). Borderline personality pathology and physical health: The role of employment. *Personality Disorders, Theory, Research, and Treatment, 9*(1), 73–80. https://doi.org/10.1037/per0000211

Cruitt, P. J., & Oltmanns, T. F. (2018a). Age-related outcomes associated with personality pathology in later life. *Current Opinion in Psychology, 21*, 89–93. https://doi.org/10.1016/j.copsyc.2017.09.013

Cruitt, P. J., & Oltmanns, T. F. (2018b). Incremental validity of self- and informant report of personality disorders in later life. *Assessment, 25*(3), 324–335. https://doi.org/10.1177/1073191117706020

Curtis, R. G., Windsor, T. D., & Soubelet, A. (2015). The relationship between Big-5 personality traits and cognitive ability in older adults – A review. *Aging, Neuropsychology, and Cognition, 22*(1), 42–71. https://doi.org/10.1080/13825585.2014.888392

Debast, I., Rossi, G., Van Alphen, S. P. J., Pauwels, E., Claes, L., Dierckx, E., … Schotte, C. K. W. (2015). Age neutrality of categorically and dimensionally measured DSM-5 section II personality disorder symptoms. *Journal of Personality Assessment, 97*(4), 321–329. https://doi.org/10.1080/00223891.2015.1021814

Dixon-Gordon, K. L., Whalen, D. J., Layden, B. K., & Chapman, A. L. (2015). A systematic review of personality disorders and health outcomes. *Canadian Psychology/Psychologie Canadienne*. https://doi.org/10.1037/cap0000024

Dondu, A., Sevincoka, L., Akyol, A., & Tataroglu, C. (2015). Is obsessive-compulsive symptomatology a risk factor for Alzheimer-type dementia? *Psychiatry Research, 225*(3), 381–386. https://doi.org/10.1016/j.psychres.2014.12.010

El-Gabalawy, R., Katz, L. Y., & Sareen, J. (2010). Comorbidity and associated severity of borderline personality disorder and physical health conditions in a nationally representative sample. *Psychosomatic Medicine, 72*(7), 641–647. https://doi.org/10.1097/PSY.0b013e3181e10c7b

Ellison, W. D., Rosenstein, L., Chelminski, I., Dalrymple, K., & Zimmerman, M. (2016). The clinical significance of single features of borderline personality disorder: Anger, affective instability, impulsivity, and chronic emptiness in psychiatric outpatients. *Journal of Personality Disorders, 30*(2), 261–270. https://doi.org/10.1521/pedi_2015_29_193

Fok, M. L. Y., Hayes, R. D., Chang, C. K., Stewart, R., Callard, F. J., & Moran, P. (2012). Life expectancy at birth and all-cause mortality among people with personality disorder. *Journal of Psychosomatic Research*. https://doi.org/10.1016/j.jpsychores.2012.05.001

Galione, J. N., & Oltmanns, T. F. (2013). The relationship between borderline personality disorder and major depression in later life: Acute versus temperamental symptoms. *The American Journal of Geriatric Psychiatry, 21*(8), 747–756. https://doi.org/10.1016/j.jagp.2013.01.026

Gleason, M. E. J., Powers, A. D., & Oltmanns, T. F. (2012). The enduring impact of borderline personality pathology: Risk for threatening life events in later middle-age. *Journal of Abnormal Psychology, 121*(2), 447–457. https://doi.org/10.1037/a0025564

Gleason, M. E. J., Weinstein, Y., Balsis, S., & Oltmanns, T. F. (2014). The enduring impact of maladaptive personality traits on relationship quality and health in later life. *Journal of Personality, 82*(6), 493–501. https://doi.org/10.1111/jopy.12068

Goldstein, R. B., Dawson, D. A., Chou, P., Ruan, W. J., Saha, T. D., Pickering, R. P., … Grant, B. F. (2008). Antisocial behavioral syndromes and past-year physical health among adults in the United States: Results from the national epidemiologic survey on alcohol and related conditions. *Journal of Clinical Psychiatry, 69*(3), 368–380. https://doi.org/10.4088/jcp.v69n0305

Gore, W. L., & Widiger, T. A. (2013). The DSM-5 dimensional trait model and five-factor models of general personality. *Journal of Abnormal Psychology, 122*(3), 816–821. https://doi.org/10.1037/a0032822

Greve, K. W., Curtis, K. L., & Bianchini, K. J. (2007). Diogenes syndrome: A five-year follow-up. *International Journal of Geriatric Psychiatry, 22*, 1166–1170. https://doi.org/10.1002/gps.1835

Greve, K. W., Curtis, K. L., Bianchini, K. J., & Collins, B. T. (2004). Personality disorder masquerading as dementia: A case of apparent Diogenes syndrome. *International Journal of Geriatric Psychiatry, 19*(7), 703–705. https://doi.org/10.1002/gps.1110

Harwood, D., Hawton, K., Hope, T., & Jacoby, R. (2001). Psychiatric disorder and personality factors associated with suicide in older people: A descriptive and case-control study. *International Journal of Geriatric Psychiatry, 16*(2), 155–165. https://doi.org/10.1002/1099-1166(200102)16:2<155::AID-GPS289>3.0.CO;2-0

Hellwig, S., Dykierek, P., Hellwig, B., Zwernemann, S., & Meyer, P. T. (2012). Alzheimer's disease camouflaged by histrionic personality disorder. *Neurocase, 18*(1), 75–79. https://doi.org/10.1080/13554794.2011.556125

Helmes, E., & Steward, L. (2010). The case of an aging person with borderline personality disorder and possible dementia. *International Psychogeriatrics/IPA, 22*(5), 840–843. https://doi.org/10.1017/S1041610210000219

Hershey, D. A., & Henkens, K. (2014). Impact of different types of retirement transitions on perceived satisfaction with life. *Gerontologist, 54*(2), 232–244. https://doi.org/10.1093/geront/gnt006

Hertzog, C., Kramer, A. F., Wilson, R. S., & Lindenberger, U. (2009). Enrichment effects on adult cognitive development. *Psychological Science in the Public Interest, 9*(1), 1–66. https://doi.org/10.1111/j.1539-6053.2009.01034.x

Hutsebaut, J., Videler, A. C., Verheul, R., & Van Alphen, S. P. J. (2019). Managing borderline personality disorder from a life course perspective: Clinical staging and health management. *Personality Disorders: Theory, Research, and Treatment, 10*(4), 309–316. https://doi.org/10.1037/per0000341

Hyde, M., Magnusson Hanson, L., Chungkham, H. S., Leineweber, C., & Westerlund, H. (2015). The impact of involuntary exit from employment in later life on the risk of major depression and being prescribed antidepressant medication. *Aging & Mental Health, 19*(5), 381–389. https://doi.org/10.1080/13607863.2014.927821

Iacovino, J. M., Powers, A. D., & Oltmanns, T. F. (2014). Impulsivity mediates the association between borderline personality pathology and body mass index. *Personality and Individual Differences, 56*, 100–104. https://doi.org/10.1016/j.paid.2013.08.028

Infurna, F. J., & Wiest, M. (2018). The effect of disability onset across the adult life span. *The Journals of Gerontology, Series B: Psychological Sciences and Social Sciences, 73*(5), 755–766. https://doi.org/10.1093/geronb/gbw055

Jahn, D. R., Poindexter, E. K., & Cukrowicz, K. C. (2015). Personality disorder traits, risk factors, and suicide ideation among older adults. *International Psychogeriatrics, 27*(11), 1785–1794. https://doi.org/10.1017/S1041610215000174

Katana, M., & Hill, P. L. (this volume). Affective aging on different time-scales. In P. L. Hill & M. Allemand (Eds.), *Personality and healthy aging in adulthood: New directions and techniques*. Cham: Springer.

Kawachi, I., & Berkman, L. F. (2001). Social ties and mental health. *Journal of Urban Health: Bulletin of the New York Academy of Medicine, 78*(3), 458–467. https://doi.org/10.1093/jurban/78.3.458

Keuroghlian, A. S., & Zanarini, M. C. (2015). Lessons learned from longitudinal studies of personality disorders. In S. K. Huprich (Ed.), *Personality disorders: Toward theoretical and empirical integration in diagnosis and assessment* (pp. 145–161). Washington DC: American Psychological Association. https://doi.org/10.1037/14549-007

Khasho, D. A., Van Alphen, S. P. J., Heijnen-Kohl, S. M. J., Ouwens, M. A., Arntz, A., & Videler, A. C. (2019). The effectiveness of individual schema therapy in older adults with borderline personality disorder: Protocol of a multiple-baseline study. *Contemporary Clinical Trials Communications, 14*, 100330. https://doi.org/10.1016/j.conctc.2019.100330

Knudsen, A. K., Skogen, J. C., Harvey, S. B., Stewart, R., Hotopf, M., & Moran, P. (2012). Personality disorders, common mental disorders and receipt of disability benefits: Evidence from the British national survey of psychiatric morbidity. *Psychological Medicine, 42*(12), 2631–2640. https://doi.org/10.1017/S0033291712000906

Korkeila, J., Oksanen, T., Virtanen, M., Salo, P., Nabi, H., Pentti, J., … Kivimä Ki, M. (2010). Early retirement from work among employees with a diagnosis of personality disorder compared to anxiety and depressive disorders. *European Psychiatry, 26*, 18–22. https://doi.org/10.1016/j.eurpsy.2009.12.022

Krueger, R. F., Hopwood, C. J., Wright, A. G. C., & Markon, K. E. (2014). *DSM-5* and the path toward empirically based and clinically useful conceptualization of personality and psychopathology. *Clinical Psychology: Science and Practice, 21*(3), 245–261. https://doi.org/10.1111/cpsp.12073

Krueger, R. F., & Markon, K. E. (2014). The role of the *DSM-5* personality trait model in moving toward a quantitative and empirically based approach to classifying personality and psychopathology. *Annual Review of Clinical Psychology, 10*, 477–501. https://doi.org/10.1146/annurev-clinpsy-032813-153732

Lee, H. B., Bienvenu, O. J., Cho, S.-J., Ramsey, C. M., Bandeen-Roche, K., Eaton, W. W., & Nestadt, G. (2010). Personality disorders and traits as predictors of incident cardiovascular disease: Findings from the 23-year follow-up of the Baltimore ECA study. *Psychosomatics, 51*(4), 289–296. https://doi.org/10.1176/appi.psy.51.4.289

Linehan, M. M. (1993). *Cognitive-behavioral treatment of borderline personality disorder*. New York, NY: Guilford Press.

Low, L.-F., Harrison, F., & Lackersteen, S. M. (2013). Does personality affect risk for dementia? A systematic review and meta-analysis. *The American Journal of Geriatric Psychiatry, 21*(8), 713–728. https://doi.org/10.1016/j.jagp.2012.08.004

Luchetti, M., Terracciano, A., Stephan, Y., & Sutin, A. R. (2016). Personality and cognitive decline in older adults: Data from a longitudinal sample and meta-analysis. *The Journals of Gerontology, Series B: Psychological Sciences and Social Sciences, 71*(4), 591–601. https://doi.org/10.1093/geronb/gbu184

Luong, G., Charles, S. T., & Fingerman, K. L. (2011). Better with age: Social relationships across adulthood. *Journal of Social and Personal Relationships, 28*(1), 9–23. https://doi.org/10.1177/0265407510391362

Lynch, T. R., Cheavens, J. S., Cukrowicz, K. C., Thorp, S. R., Bronner, L., & Beyer, J. (2007). Treatment of older adults with co-morbid personality disorder and depression: A dialectical behavior therapy approach. *International Journal of Geriatric Psychiatry, 22*, 131–143. https://doi.org/10.1002/gps.1703

McDavid, J. D., & Pilkonis, P. A. (1996). The stability of personality disorder diagnoses. *Journal of Personality Disorders, 10*(1), 1–15. https://doi.org/10.1521/pedi.1996.10.1.1

McWilliams, L. A., Clara, I. P., Murphy, P. D. J., Cox, B. J., & Sareen, J. (2008). Associations between arthritis and a broad range of psychiatric disorders: Findings from a nationally representative sample. *The Journal of Pain, 9*(1), 37–44. https://doi.org/10.1016/j.jpain.2007.08.002

Morse, J. Q., Pilkonis, P. A., Houck, P. R., Frank, E., & Reynolds, C. F. (2005). Impact of cluster C personality disorders on outcomes of acute and maintanence treatment in late-life depression. *The American Journal of Geriatric Psychiatry, 13*(9), 808–814. https://doi.org/10.1176/appi.ajgp.13.9.808

Mosca, I., & Barrett, A. (2016). The impact of voluntary and involuntary retirement on mental health: Evidence from older Irish adults. *Journal of Mental Health Policy and Economics, 19*(1), 33–44.

Newton-Howes, G., Clark, L. A., & Chanen, A. (2015). Personality disorder across the life course. *The Lancet, 385*(9969), 727–734. https://doi.org/10.1016/S0140-6736(14)61283-6

Nicholas, H., Moran, P., Foy, C., Brown, R. G., Lovestone, S., Bryant, S., & Boothby, H. (2010). Are abnormal premorbid personality traits associated with Alzheimer's disease? - A case-control study. *International Journal of Geriatric Psychiatry, 25*, 345–351. https://doi.org/10.1002/gps.2345

Nock, M. K., Borges, G., Bromet, E. J., Cha, C. B., Kessler, R. C., & Lee, S. (2008). Suicide and suicidal behavior. *Epidemiologic Reviews, 30*(1), 133–154. https://doi.org/10.1093/epirev/mxn002

Nordentoft, M., Wahlbeck, K., Hällgren, J., Westman, J., Ösby, U., Alinaghizadeh, H., … Laursen, T. M. (2013). Excess mortality, causes of death and life expectancy in 270,770 patients with recent onset of mental disorders in Denmark, Finland and Sweden. *PLoS One, 8*(1). https://doi.org/10.1371/journal.pone.0055176

Oltmanns, T. F., & Balsis, S. (2011). Personality disorders in later life: Questions about the measurement, course, and impact of disorders. *Annual Review of Clinical Psychology, 7*, 321–349. https://doi.org/10.1146/annurev-clinpsy-090310-120435

Oltmanns, T. F., Rodrigues, M. M., Weinstein, Y., & Gleason, M. E. J. (2014). Prevalence of personality disorders at midlife in a community sample: Disorders and symptoms reflected in interview, self, and informant reports. *Journal of Psychopathology and Behavioral Assessment, 36*(2), 177–188. https://doi.org/10.1007/s10862-013-9389-7

Paris, J. (2003). *Personality disorders over time: Precursors, course and outcome*. Quebec, Canda: American Psychiatric Publishing.

Pilleron, S., Clément, J.-P., Ndamba-Bandzouzi, B., Mbelesso, P., Dartigues, J.-F., Preux, P.-M., … Lambert, J.-C. (2015). Is dependent personality disorder associated with mild cognitive impairment and dementia in Central Africa? A result from the EPIDEMCA programme. *International Psychogeriatrics, 27*(2), 279–288. https://doi.org/10.1017/S104161021400180X

Powers, A. D., Gleason, M. E. J., & Oltmanns, T. F. (2013). Symptoms of borderline personality disorder predict interpersonal (but not independent) stressful life events in a community sample of older adults. *Journal of Abnormal Psychology, 122*(2), 469–474. https://doi.org/10.1037/a0032363

Powers, A. D., & Oltmanns, T. F. (2012). Personality disorders and physIcal health: A longitudinal examination of physical functioning, healthcare utilization, and health-related behaviors in middle-aged adults. *Journal of Personality Disorders, 26*(4), 524–538. https://doi.org/10.1521/pedi.2012.26.4.524

Powers, A. D., & Oltmanns, T. F. (2013). Borderline personality pathology and chronic health problems in later adulthood: The mediating role of obesity. *Personality Disorders, Theory, Research, and Treatment, 4*(2), 152–159. https://doi.org/10.1037/a0028709

Powers, A. D., Strube, M. J., & Oltmanns, T. F. (2014). Personality pathology and increased use of medical resources in later adulthood. *The American Journal of Geriatric Psychiatry, 22*, 1478–1486. https://doi.org/10.1016/j.jagp.2013.10.009

Prior, J., Abraham, R., Nicholas, H., Chan, T., Vanvlymen, J., Lovestone, S., & Boothby, H. (2016). Are premorbid abnormal personality traits associated with behavioural and psychological symptoms in dementia? *International Journal of Geriatric Psychiatry, 31*(9), 1050–1055. https://doi.org/10.1002/gps.4418

Quirk, S. E., Berk, M., Chanen, A. M., Koivumaa-Honkanen, H., Brennan-Olsen, S. L., Pasco, J. A., & Williams, L. J. (2016). Population prevalence of personality disorder and associations with physical health comorbidities and health care service utilization: A review. *Personality Disorders, Theory, Research, and Treatment, 7*(2), 136–146. https://doi.org/10.1037/per0000148

Rosowsky, E., Lodish, E., Ellison, J. M., & Van Alphen, S. P. J. (2019). A Delphi study of late-onset personality disorders. *International Psychogeriatrics*, 10–16. https://doi.org/10.1017/S1041610218001473

Samuels, J., Eaton, W. W., Bienvenu, O. J., Brown, C. H., Costa, P. T., & Nestadt, G. (2002). Prevalence and correlates of personality disorders in a community sample. *The British Journal of Psychiatry, 180*, 536–542. https://doi.org/10.1192/bjp.180.6.536

Schuster, J.-P., Hoertel, N., Le Strat, Y., Manetti, A., & Limosin, F. (2013). Personality disorders in older adults: Findings from the national epidemiologic survey on alcohol and related conditions. *The American Journal of Geriatric Psychiatry, 21*(8), 757–768. https://doi.org/10.1016/j.jagp.2013.01.055

Segal, D. L., Marty, M. A., Meyer, W. J., & Coolidge, F. L. (2012). Personality, suicidal ideation, and reasons for living among older adults. *The Journals of Gerontology, Series B: Psychological Sciences and Social Sciences, 67*(2), 159–166. https://doi.org/10.1093/geronb/gbr080

Shah, A., Bhat, R., Zarate-Escudero, S., Deleo, D., & Erlangsen, A. (2016). Suicide rates in five-year age-bands after the age of 60 years: The international landscape. *Aging and Mental Health, 20*(2), 131–138. https://doi.org/10.1080/13607863.2015.1055552

Skodol, A. E., Pagano, M. E., Bender, D. S., Shea, M. T., Gunderson, J. G., Yen, S., … McGlashan, T. H. (2005). Stability of functional impairment in patients with schizotypal, borderline, avoidant, or obsessive-compulsive personality disorder over two years. *Psychological Medicine, 35*(3), 443–451. https://doi.org/10.1017/S003329170400354X

Skodol, A. E., Stout, R. L., McGlashan, T. H., Grilo, C. M., Gunderson, J. G., Trade Shea, M., … Oldham, J. M. (1999). Co-occurrence of mood and personality disorders: A report from the collaborative longitudinal personality disorders study (CLPS). *Depression and Anxiety, 10*(4), 175–182. https://doi.org/10.1002/(SICI)1520-6394(1999)10:4<175::AID-DA6>3.0.CO;2-2

Sleep, C. E., Lamkin, J., Lynam, D. R., Campbell, W. K., & Miller, J. D. (2018). Personality disorder traits: Testing insight regarding presence of traits, impairment, and desire for change. *Personality Disorders, Theory, Research, and Treatment, 10*(2), 123–131. https://doi.org/10.1037/per0000305

Smith, G. E. (2016). Healthy cognitive aging and dementia prevention. *American Psychologist, 71*(4), 268–275. https://doi.org/10.1037/a0040250

Suzuki, T., Samuel, D. B., Pahlen, S., & Krueger, R. F. (2015). *DSM-5* alternative personality disorder model traits as maladaptive extreme variants of the five-factor model: An item-response theory analysis. *Journal of Abnormal Psychology, 124*(2), 343–354. https://doi.org/10.1037/abn0000035

Terracciano, A., Stephan, Y., Luchetti, M., Albanese, E., & Sutin, A. R. (2017). Personality traits and risk of cognitive impairment and dementia. *Journal of Psychiatric Research, 89*, 22–27. https://doi.org/10.1016/j.jpsychires.2017.01.011

Terracciano, A., Sutin, A. R., An, Y., Brien, R. J. O., Ferrucci, L., Zonderman, A. B., & Resnick, S. M. (2014). Personality and risk of Alzheimer's disease: New data and meta-analysis. *Alzheimer's & Dementia: The Journal of the Alzheimer's Association, 10*(2), 179–186. https://doi.org/10.1016/j.jalz.2013.03.002

Trull, T. J., Jahng, S., Tomko, R. L., Wood, P. K., & Sher, K. J. (2010). Revised NESARC personality disorder diagnoses: Gender, prevalence, and comorbidity with substance dependence disorders. *Journal of Personality Disorders, 24*(4), 412–426. https://doi.org/10.1521/pedi.2010.24.4.412

Tyrer, P., Crawford, M., Mulder, R., Blashfield, R., Farnam, A., Fossati, A., … Reed, G. M. (2011). The rationale for the reclassification of personality disorder in the 11th revision of the international classification of diseases (ICD-11). *Personality and Mental Health, 5*, 256–259. https://doi.org/10.1002/pmh.190

Van Alphen, S. P. J., Bolwerk, N., Videler, A. C., Tummers, J. H. A., van Royen, R. J. J., Barendse, H. P. J., … Rosowsky, E. (2012). Age-related aspects and clinical implications of diagnosis and treatment of personality disorders in older adults. *Clinical Gerontologist, 35*(1), 27–41. https://doi.org/10.1080/07317115.2011.628368

Van Alphen, S. P. J., Van Dijk, S. D. M., Videler, A. C., Rossi, G., Dierckx, E., Bouckaert, F., & Oude Voshaar, R. C. (2015). Personality disorders in older adults: Emerging research issues. *Current Psychiatry Reports, 17*, 538. https://doi.org/10.1007/s11920-014-0538-9

Van Den Broeck, J., Bastiaansen, L., Rossi, G., Dierckx, E., & De Clercq, B. (2013). Age-neutrality of the trait facets proposed for personality disorders in DSM-5: A DIFAS analysis of the PID-5. *Journal of Psychopathology and Behavioral Assessment, 35*(4), 487–494. https://doi.org/10.1007/s10862-013-9364-3

Van Den Broeck, J., Bastiaansen, L., Rossi, G., Dierckx, E., De Clercq, B., & Hofmans, J. (2014). Hierarchical structure of maladaptive personality traits in older adults: Joint factor analysis of the PID-5 and the DAPP-BQ. *Journal of Personality Disorders, 28*(2), 198–211. https://doi.org/10.1521/pedi_2013_27_114

van Solinge, H. (2007). Health change in retirement. *Research on Aging, 29*(3), 225–256. https://doi.org/10.1177/0164027506298223

Verheul, R., Bartak, A., & Widiger, T. (2007). Prevalence and construct validity of personality disorder not otherwise specified (PDNOS). *Journal of Personality Disorders, 21*(4), 359–370. https://doi.org/10.1521/pedi.2007.21.4.359

Verheul, R., & Widiger, T. A. (2004). A meta-analysis of the prevalence and usage of the personality disorder not otherwise specified (PDNOS) diagnosis. *Journal of Personality Disorders, 18*(4), 309–319. https://doi.org/10.1521/pedi.2004.18.4.309

Videler, A. C., Hutsebaut, J., Schulkens, J. E. M., Sobczak, S., & Van Alphen, S. P. J. (2019). A life span perspective on borderline personality disorder. *Current Psychiatry Reports, 21*(51), 1–8. https://doi.org/10.1007/s11920-019-1040-1

Videler, A. C., Rossi, G., Schoevaars, M., Van Der Feltz-Cornelis, C. M., & Van Alphen, S. P. J. (2014). Effects of schema group therapy in older outpatients: A proof of concept study. *International Psychogeriatrics, 26*(10), 1709–1717. https://doi.org/10.1017/S1041610214001264

Videler, A. C., Van der Feltz-Cornelis, C. M., Rossi, G., Van Royen, R. J. J., Rosowsky, E., & Van Alphen, S. P. J. (2015). Psychotherapeutic treatment levels for personality disorders in older adults. *Clinical Gerontologist, 38*(4), 325–341. https://doi.org/10.1080/07317115.2015.1032464

World Health Organization. (2018). International statistical classification of diseases and related health problems. Retrieved from https://icd.who.int/browse11/l-m/en

Zanarini, M. C., Jacoby, R. J., Frankenburg, F. R., Reich, D. B., & Fitzmaurice, G. (2009). The 10-year course of social security disability income reported by patients with borderline personality disorder and axis II comparison subjects. *Journal of Personality Disorders, 23*(4), 346–356. https://doi.org/10.1521/pedi.2009.23.4.346

Zimmerman, M., Rothschild, L., & Chelminski, I. (2005). The prevalence of DSM-IV personality disorders in psychiatric outpatients. *The American Journal of Psychiatry, 162*(10), 1911–1918. https://doi.org/10.1176/appi.ajp.162.10.1911

Chapter 5
Affective Aging on Different Time-Scales

Marko Katana and Patrick L. Hill

5.1 Introduction

With rising life expectancy, the importance of healthy aging into older adulthood has become increasingly relevant. However, gerontology and lifespan psychology have struggled to find a commonly accepted set of criteria to define healthy aging. Lifespan theories have suggested that individuals may maintain psychological functioning by selectively choosing contexts, optimizing available resources, and compensating for age-related losses (e.g., Freund & Baltes, 2002). One theoretical framework argues that individuals may maintain affective stability by increasing their secondary control strategies, such as actively shaping their environment to compensate for losses (Heckhausen, 1995; Heckhausen, Wrosch, & Schulz, 2010; Schulz & Heckhausen, 1996). Yet another, albeit related, theoretical framework argues that individuals are effective in maintaining a positive view by decreasing tenacious goal pursuit and increasing flexible goal adjustment with old age (Brandtstädter, Wentura, & Greve, 1993). In other words, individuals actively shape their own lives and environments within the potentials and limits given by individual, social, cultural, and biological constraints (Freund & Riediger, 2003). Building from these theories, successful aging may be best described as a dynamic interaction between individuals and their environment. In line with these prominent lifespan theories, the most recent model of healthy aging by the World Health

M. Katana (✉)
Department of Psychology and University Research Priority Program "Dynamics of Healthy Aging", University of Zurich, Zürich, Switzerland
e-mail: marko.katana@uzh.ch

P. L. Hill
Department of Psychological and Brain Sciences, Washington University in St. Louis, St. Louis, MO, USA
e-mail: patrick.hill@wustl.edu

P. L. Hill, M. Allemand (eds.), *Personality and Healthy Aging in Adulthood*,
International Perspectives on Aging 26,
https://doi.org/10.1007/978-3-030-32053-9_5

Organization does not define a specific threshold of functioning or health as key to healthy aging. Instead, it argues that healthy aging is an ongoing process between individuals' abilities, including intrinsic capacities (composite of all physical and mental capacities of an individual) and environmental factors (World Health Organization, 2015).

The World Health Organization model also emphasizes the need to move beyond physical health, and for researchers to consider more subjective aspects such as personal well-being as indicators of (and potentially catalysts for) healthy aging. Subjective well-being includes an affective component such as pleasant and unpleasant affect, and a cognitive component such as domain specific satisfaction and global judgments of life satisfaction (Diener, Suh, Lucas, & Smith, 1999). The measurement of the affective component of well-being is, however, still being debated in the current literature. Some argue that affective well-being is best measured as a bipolar construct with positive affect opposed to negative affect (Feldman Barrett & Russell, 1998; Russell, 2017). Others argue that affective well-being is a bi-dimensional construct with positive and negative affect as two independent factors (Cacioppo & Berntson, 1994; Larsen, 2017; Watson, Clark, & Tellegen, 1988). In this overview, we present the research findings for both positive and negative affect as separate constructs, allowing the reader to decide whether to consider them as bipolar ends or more independent constructs.

5.2 Age Differences and Long-Term Change of Affect

Prior to considering how affect changes with aging, it first is worth orienting the reader to a discussion of how affective constructs operate within a temporal hierarchy. One framework for considering this hierarchy suggests that affective traits are viewed as the affect construct with the longest duration or consistency, are more dispositional in nature, and they are often assessed using self-reports that capture how someone typically is (Rosenberg, 1998). On the other end of the spectrum are emotions, which are more ephemeral in nature, sometimes lasting just seconds, and thus must be captured in the moment (Ekman, 1984). The in-between level often is labeled "moods," which can last for hours or days. This chapter primarily will discuss affective traits and emotions, starting with age differences in affect at the trait level (as most research has considered developmental trends at this level).

At the trait level, work suggests relative stability for affective well-being for most of the adult lifespan (Charles & Carstensen, 2007; Scheibe & Carstensen, 2010), though meta-analytic work has suggested the potential for greater declines in positive affect and increases in negative affect during later adulthood (Pinquart, 2001). However, patterns seem to differ based on affective arousal, defined as the intensity or extent of affective activation; for example, high-arousal negative affect occurs less frequently and less intensely in older adults whereas positive affect is characterized by high stability across adulthood into old age (Kunzmann, Kappes, & Wrosch, 2014).

One explanation for these patterns comes from how individuals change their emotion-focused goals during adulthood. For instance, the Socioemotional Selectivity Theory (SST; Carstensen, 2006; Carstensen, Isaacowitz, & Charles, 1999) discusses the relevance of informational and emotional goal changes as a function of future time perception. Individuals with an open-ended time perception favor informational and knowledge-based goals, often with the intent of learning more about oneself. In contrast, individuals with a limited time perception who favor emotional goals, insofar that they prefer to prioritize their affective well-being over other potential pursuits. Given that older adults tend to hold a more limited perception, adults tend to focus more on emotional goals rather than information seeking in older adulthood.

Another line of research suggests that problems in emotion regulation may arise with a late life decline of cognitive resources. The Dynamic Integration Theory (DIT) (Labouvie-Vief, 2003) posits that this decline in cognitive capacities in older age may be associated with a decreased ability to flexibly respond to the demands of situations and therefore older adults report less effective emotion regulation (Scott, Sliwinski, & Blanchard-Fields, 2013). As a result of age-related declines, older adults report greater difficulties to cognitively tolerate negative affect and thus selectively choose situations with less probability for negative affect. In other words, older adults are less effective than younger adults at regulating their affective experience when situational demands exceed their available cognitive capacities (Wrzus, Müller, Wagner, Lindenberger, & Riediger, 2013).

These two contemporary lifespan theories (SST and DIT) on affective functioning seem to form opposing predictions on the underlying processes of affective aging. An influential extension of these two approaches is the Strength and Vulnerability Integration (SAVI) model (Charles, 2010). In line with SST, this model posits aging to be associated with changes in time perception which lead to an improvement of emotional competencies due to arguably better emotion regulation strategies and an attentional shift to avoid negative stimuli. However, SAVI argues that aging is also associated with reduced physiological flexibility to contextual change, which makes older adults more vulnerable. The vulnerabilities described in SAVI form a complement to the cognitive decline in affective cognitive complexity described in DIT. Affective well-being is considered to be the a result of these emotional strengths as well as physiological and cognitive vulnerabilities (Charles, 2010).

5.3 Affect on Different Time-Scales

Studies based on the theories discussed above have mainly focused on between-person designs, even though these theories often emphasize within-person regulatory processes in their framing. Indeed, affect does not only show variation between individuals but also typically varies across situations within individuals. From a within-person perspective, it is valuable to consider at least two time-scales. First, long-term intraindividual change characterizes variation that is indicative of development expected to take place on a macro-level timescale over years or months. Second, short-term intraindividual variability represents within-person variation on a micro-level timescale assessed daily or multiple times per day. Using these assessments, intraindividual variability of affect can be measured using individual standard deviations over daily assessments (Eid & Diener, 1999). In the case of affective well-being, relatively stable pattern of affective experiences over years is referred to as an affective trait. However, momentary affect and emotions are more situation-specific and therefore vary in much shorter time intervals.

Focusing on intraindividual variability, one can further distinguish between net intraindividual variability and time-structured intraindividual variability (for a review, Ram & Gerstorf, 2009; Röcke & Brose, 2013). The univariate net variability approach describes the mere amount of affect variability such as the magnitude of fluctuations without taking temporal dependencies into account (e.g., intraindividual standard deviation, iSD). The multivariate net variability approach includes multiple time-varying variables and describes the correlated within-person variation. Contemporaneous within-person couplings can be operationalized by multilevel regression coefficients or P-technique (Ong & Bergeman, 2004). Notably, these approaches are still correlational and do not allow interpretation about causality.

In contrast to the net variability approaches, the time-structured approaches allow interpretations about causality to a certain extent and an even better glimpse into processes. The univariate time-structured approach includes time as a relevant metric in the analyses in order to study the duration of affective episodes or the temporal return to "baseline." Differential equation models such as the damped linear oscillator model (Boker, 2013), dynamics of affect model (Kuppens, Oravecz, & Tuerlinckx, 2010) or dynamic structural equation models (Hamaker, Asparouhov, Brose, Schmiedek, & Muthén, 2017) have the potential to model multiple aspects of time-varying affect simultaneously (e.g., inertia, attractor strength). A comprehensive discussion of these models is beyond the reach of the current chapter. However, we encourage the reader to seek out these references for additional information, given that the use of multivariate within-person dynamics is still sparse in psychology (though for an example study, see Scott et al., 2013).

5.4 Research on Affect in Laboratory and in Real Life

In recent years, a number of researchers have emphasized the need for collecting high-frequency daily life data in order to study intraindividual associations of psychological functioning in participants' natural context in addition to traditional laboratory studies (Brose & Ebner-Priemer, 2015; Hamaker, 2012; Hoppmann & Riediger, 2009). The study of intraindividual variability of affect is well-suited for high frequency assessment as affective states can vary rapidly from time point to time point and are sensitive to contextual change (Brose & Ebner-Priemer, 2015). Nevertheless, reactions to realistic contextual change often cannot be studied in laboratories without limitations. Notably, laboratory studies isolate participants from their everyday lives such as routine physical activities, social interactions and context specific motivational goals (Hamaker, 2012). Therefore, there is a need for studies in real life that can take contextual information into account to understand emotional experience in more depth and in a more naturalistic way.

Compared to traditional laboratory studies, everyday life studies have four main advantages. First, these studies often have greater ecological validity, insofar that it holds greater accuracy for representing the real world. Second, momentary self-report minimizes the retrospective biases that are more likely in traditional self-report. Third, micro-longitudinal observations allow to study short-term within-person processes, such as evaluating changes in behavior and psychological functioning. Fourth, studies in a naturalistic setting facilitate the assessment of contextual information, which enables researchers to link situational aspects to person-specific variations (Brose & Ebner-Priemer, 2015; Hoppmann & Riediger, 2009).

Due to technological progress, assessing data regularly in real-life has become increasingly possible. In so doing, researchers have the ability to study the dynamic nature of behavior and psychological functioning within a person over extended periods of time and with high control over the timing of the responses. These everyday life measurements can include self-reports, physiological or biological data as well as observed behaviors over a predefined period of time assessed via passive sensing or active responses and task performance (Ram & Gerstorf, 2009; Trull & Ebner-Priemer, 2013).

Recommendations for Research in Daily Life To truly capture intraindividual variability in affect, study designs should apply an experience sampling approach by using questionnaires on participants' smartphones, asking participants to rate their momentary affect. Depending on the variable observed, the length of the study and the frequency of assessments are crucial to capture dynamic processes. Too small of intervals between assessments might lead to the misperception of stability, while sampling with too large of intervals might miss the actual fluctuation (Boker & Nesselroade, 2002; Ram & Gerstorf, 2009). For the assessment of affect, a higher frequency per day (e.g., morning and evening or more) can give a more fine-grained picture of how affective states unfold in everyday life. For example, such studies would make it possible to investigate whether early morning affect is associated with affect in the afternoon or in the evening, taking circadian patterns into account.

High-frequency assessments of affect allow researchers to use time-structured approaches to model processes in everyday life in more detail than net variability approaches do (Röcke & Brose, 2013). For example, zooming in on a micro timescale would allow researchers to study the duration of an emotional episode or the temporal return to the individual's usual affect (Kuppens et al., 2010). However, the question regarding what would be an appropriate time interval to catch fluctuations in affect in everyday life remains to be answered. It is likely that methods assess affect retrospectively at the end of the day may better capture daily mood, whereas the experience sampling approach truly captures intraindividual variability in momentary affect. Notably, there are also end-of-day diaries that specifically ask participants to rate their momentary affect; however, they mostly aim at reconstructing the affective states experienced throughout the day (Röcke, Hoppmann, & Klumb, 2011).

Lastly, the interpretation of affective variability highly depends on the timescale that was chosen to be investigated. On the one hand, stability in affect over multiple days is seen as a sign of intact psychological functioning (Gruber, Kogan, Quoidbach, & Mauss, 2013). On the other hand, variability on a shorter timescale (i.e., momentary assessment) is seen as adaptive when looking at event occurrences during the day (Kashdan & Rottenberg, 2010). Therefore, longer time intervals might be better suited to study (in)stability of affective functioning, whereas shorter time intervals are better suited to study the adaptivity of affect to contextual changes.

However, it is important to bear in mind that these studies also hold some disadvantages. Participants who are aware of being tracked by mobile sensors might behave differently than in an unobserved condition. For example, individuals might show greater movement and higher activity when being tracked (Duncan, Badland, & Mummery, 2009). Furthermore, study compliance may change across a study's time span (Trull & Ebner-Priemer, 2013). It would be reasonable to expect a drop in compliance as participants start to lose interest, or forget to recharge their devices on a regular basis. These factors depend on the population being studied and both reactivity and compliance can be optimized with design choices that take care in minimizing participants' burden to the extent possible given a particular research focus (Trull & Ebner-Priemer, 2013).

5.5 Taking the Context into Account for Affective Aging

The value of moving beyond trait-level assessments of affect allows investigations into how older and younger adults are differentially influenced by the given context or daily environment. Event occurrences, such as daily hassles and uplifts, provide valuable situations for considering the role of context on emotion in studies of everyday life (Charles et al., 2010; Scott et al., 2013; Wrzus, Luong, Wagner, & Riediger, 2015). For instance, research suggests that older adults report fewer interpersonal tensions, experienced less stress, and were less likely to argue in response to interpersonal tensions (Birditt, Fingerman, & Almeida, 2005). These results may hint at better emotion regulation in older adults or an avoidance of high arousal situations, in line with other studies that have found that older adults' emotional reactivity to daily stressors was less intense in comparison to younger adults (Brose, Schmiedek, Lövdén, & Lindenberger, 2011; Hay & Diehl, 2011). However, some work does find contradictory results or no age differences in reactivity across various everyday life domains (Mroczek & Almeida, 2004; Neupert, Almeida, & Charles, 2007; Röcke, Li, & Smith, 2009; Wrzus et al., 2013). Possible explanations for these ambiguous findings include the differences across studies in their measurements, age ranges, and time frames (Röcke, Brose, & Kuppens, 2018).

When considering the SAVI model, one would expect that older adults may experience weaker emotional flexibility in overtaxing situations and a slower return to homeostasis in situations when avoidance of emotional distress is not possible (Charles, 2010). However, emotional reactivity in form of time passed since the occurrence of the hassle did not show age differences (Scott et al., 2013). Research using multiple assessments per day though is needed in order to show a more comprehensive picture of the association between momentary hassles or momentary uplifts and affective experience.

When considering specific daily events associated with shifts in affective experience, and well as events known to promote healthy aging, physical activity and exercise provides a good case example. Physical activity is seen as a health enhancing behavior and therefore associated with emotional well-being of individuals (Schwarzer & Leppin, 1991; Schwarzer & Luszczynska, 2008). In recent years, there has been multiple studies that specifically examined the within-person association between physical activity and affect. Studies with university students have shown that assessments with higher physical activity were related to moments characterized by higher valence and arousal and lower calmness (Kanning, Ebner-Priemer, & Brand, 2012; Kanning et al., 2012).

To our knowledge, only three studies thus far have investigated the within-person association between physical activity and affective states in healthy older adults. In one study, 69 older adults aged 50–70 years participated in an ambulatory assessment study during three consecutive days. Every time a predefined threshold of physical activity or inactivity was surpassed, the participants were prompted on their smartphones afterwards to report on affect. Results showed that participants with higher physical activity reported more arousal and less calmness 10 minutes after the assessment (Kanning, Ebner-Priemer, & Schlicht, 2015). Another study with adults over 50 years also found that physical activity was positively associated with arousal and negatively associated with calmness (Kanning & Hansen, 2017). Finally, daily physical activity appears negatively associated with daily depressive mood within-persons in healthy older adults (Gruenenfelder-Steiger et al., 2017). As such, though further research is needed, it appears that physical activity may promote both healthy physical aging and positive emotional states for older adults.

5.6 Taking the Person into Account for Affective Aging

Another factor influencing healthy emotional aging is the individual's personality. Personality traits are frequently described with respect to the Big Five model (Costa & McCrae, 1992; John, Naumann, & Soto, 2008). Within this framework, it is well-established that certain personality traits are associated with trait affect, in line with the suggestion that a personality trait is comprised of cognitive, behavioral, and affective components (Roberts, 2009). With respect to the Big Five traits, extraversion and neuroticism are most reflective of positive and negative emotionality, respectively. Extraversion is defined with respect to greater energy and arousal, whereas neuroticism is viewed as a dispositional tendency to experience greater anxiety, stress, and depressive affect (Costa & McCrae, 1992; John et al., 2008). Though meta-analytic work found robust associations between these two traits and affect, it is worth noting that the other three traits show consistent, though more modest, associations with trait affect as well (Steel, Schmidt, & Shultz, 2008). For instance, conscientiousness, defined by greater industriousness and self-control, typically associates with more positive and less negative affect, while agreeableness, the tendency to be friendly and trust others, also appears associated with a more positive trait affective profile.

When considering more micro-level time scales, state affect also demonstrates associations with personality trait profiles. Of the Big Five traits, neuroticism appears the most consistent predictor of state affect; individuals higher on neuroticism tend to report more daily negative affect, more stressful experiences in daily life, and tend to have greater negative affective reactivity to those stressful events (Mroczek & Almeida, 2004). Extraversion, again, appears important for understanding daily affect, insofar that the trait predicts greater momentary positive affect, particularly when an individual acts in line with their extraverted tendencies (Fleeson, Malanos, & Achille, 2002).

Only a handful of studies have considered the associations between trait personality and affect within a lifespan developmental context. Some research though suggests that traits may be less predictive of negative affect for older than younger adults, when asked about affect in a single laboratory setting (Ready & Robinson, 2008). This work was interpreted as supportive evidence for the importance of emotion regulation strategies during older adulthood. However, longitudinal research suggests that neuroticism may actually prove a stronger negative predictor of future emotional well-being for older than younger adults (Ready, Åkerstedt, & Mroczek, 2012). Accordingly, further work is needed to understand whether the role of personality on affect differs across adulthood, particularly with respect to understanding how this trait differentially predicts affective aging across different time-scales.

5.7 Conclusion

In sum, to capture individual development over time, one must understand that person's affective dispositions and daily levels. Moreover, any effort to understand affective aging must consider the contexts of daily life, which change throughout adulthood. With shifting environmental demands, due to fluctuations in roles and commitments, researchers need to carefully consider the person-in-situation when charting changes in affective well-being over time. Moreover, it is clear that researchers need to thoroughly account for situational demands, both when studying healthy aging in general, and with respect specifically to emotional reactivity and regulation strategies. Though the recommendations from this chapter, and the methods described therein, present many challenges, they also present many opportunities for the study of personality and healthy aging. Indeed, few pursuits may be as rewarding as understanding how to help people become happier and less stressed with age.

Acknowledgments Work on the manuscript was supported by a grant from the Jacobs Foundation to the first author. During the work on his dissertation, Marko Katana was a pre-doctoral fellow of the International Max Planck Research School on the Life Course (LIFE, www.imprs-life.mpg.de; participating institutions: Max Planck Institute for Human Development, Freie Universität Berlin, Humboldt-Universität zu Berlin, University of Michigan, University of Virginia, University of Zurich). Thank you to Christina Röcke for her feedback on earlier versions of this manuscript.

References

Birditt, K. S., Fingerman, K. L., & Almeida, D. M. (2005). Age differences in exposure and reactions to interpersonal tensions: A daily diary study. *Psychology and Aging, 20*(2), 330–340. https://doi.org/10.1037/0882-7974.20.2.330

Boker, S. M. (2013). Selection, optimization, compensation, and equilibrium dynamics. *Journal of Gerontopsychology and Geriatric Psychiatry, 29*, 61–73. Retrieved from http://people.virginia

Boker, S. M., & Nesselroade, J. R. (2002). A method for modeling the intrinsic dynamics of intraindividual variability: Recovering the parameters of simulated oscillators in multi-wave panel data. *Multivariate Behavioral Research, 37*, 127–160. https://doi.org/10.1207/S15327906MBR3701_06

Brandtstädter, J., Wentura, D., & Greve, W. (1993). Adaptive resources of the aging self: Outlines of an emergent perspective. *International Journal of Behavioral Development, 16*(2), 323–349.

Brose, A., & Ebner-Priemer, U. W. (2015). Ambulatory assessment in the research on aging: Contemporary and future applications. *Gerontology, 61*, 372–380. https://doi.org/10.1159/000371707

Brose, A., Schmiedek, F., Lövdén, M., & Lindenberger, U. (2011). Normal aging dampens the link between intrusive thoughts and negative affect in reaction to daily Stressors. *Psychology and Aging, 26*(2), 488–502. https://doi.org/10.1037/a0022287

Cacioppo, J. T., & Berntson, G. G. (1994). Relationship between attitudes and evaluative space: A critical review, with emphasis on the separability of positive and negative substrates. *Psychological Bulletin, 115*(3), 401–423. https://doi.org/10.1037//0033-2909.115.3.401

Carstensen, L. L. (2006). The influence of a sense of time on human development. *Science, 312*, 1913–1915. https://doi.org/10.1126/science.1127488

Carstensen, L. L., Isaacowitz, D. M., & Charles, S. T. (1999). Taking time seriously: A theory of socioemotional selectivity. *American Psychologist, 54*, 165–181. https://doi.org/10.1037//0003-066X.54.3.165

Charles, S. T. (2010). Strength and vulnerability integration: A model of emotional well-being across adulthood. *Psychological Bulletin, 136*, 1068–1091. https://doi.org/10.1037/a0021232

Charles, S. T., & Carstensen, L. L. (2007). Emotion regulation and aging. In J. J. Gross (Ed.), *Handbook of emotion regulation* (pp. 307–320). New York: Guilford.

Charles, S. T., Luong, G., Almeida, D. M., Ryff, C., Sturm, M., & Love, G. (2010). Fewer ups and downs: Daily stressors mediate age differences in negative affect. *Journal of Gerontology: Psychological Sciences, 65B*, 279–286. https://doi.org/10.1093/geronb/gbq002

Costa, P. T., & McCrae, R. R. (1992). Four ways five factors are basic. *Personality and Individual Differences, 13*, 653–665.

Diener, E., Suh, E. M., Lucas, R. E., & Smith, H. L. (1999). Subjective well-being: Three decades of progress. *Psychological Bulletin, 125*, 276–302. https://doi.org/10.1037/0033-2909.125.2.276

Duncan, M. J., Badland, H. M., & Mummery, W. K. (2009). Applying GPS to enhance understanding of transport-related physical activity. *Journal of Science and Medicine in Sport, 12*, 549–556. https://doi.org/10.1016/j.jsams.2008.10.010

Eid, M., & Diener, E. (1999). Intraindividual variability in affect: Reliability, validity, and personality correlates. *Journal of Personality and Social Psychology, 76*, 662–676. https://doi.org/10.1037/0022-3514.76.4.662

Ekman, P. (1984). Expression and the nature of emotion. In K. R. Scherer & P. Ekman (Eds.), *Approaches to emotion* (pp. 319–344). Hillsdale, NJ: Lawrence Erlbaum Assodates Inc..

Feldman Barrett, L., & Russell, J. A. (1998). Independence and Bipolarity in the Structure of Current Affect. *Journal of Personality and Social Psychology, 74*(4), 967–984. Retrieved from https://www.homeworkforyou.com/static/uploadedfiles/User_2768452013psp-74-4-967.pdf

Fleeson, W., Malanos, A. B., & Achille, N. M. (2002). An intraindividual process approach to the relationship between extraversion and positive affect: Is acting extraverted as "good" as being extraverted? *Journal of Personality and Social Psychology, 83*, 1409–1422. https://doi.org/10.1037//0022-3514.83.6.1409

Freund, A. M., & Baltes, P. B. (2002). Life-management strategies of selection, optimization, and compensation: Measurement by self-report and construct validity. *Journal of Personality and Social Psychology, 82*(4), 642–662. https://doi.org/10.1037/0882-7974.13.4.531

Freund, A. M., & Riediger, M. (2003). Successful aging. In R. M. Lerner, M. A. Easterbrooks, & J. Mistry (Eds.), *Handbook of psychology* (pp. 601–628). New York, NY: Wiley.

Gruber, J., Kogan, A., Quoidbach, J., & Mauss, I. B. (2013). Happiness is best kept stable: Positive emotion variability is associated with poorer psychological health. *Emotion, 1*, 1–6. https://doi.org/10.1037/a0030262

Gruenenfelder-Steiger, A. E., Katana, M., Martin, A. A., Aschwanden, D., Koska, J. L., Kündig, Y., … Allemand, M. (2017). Physical activity and depressive mood in the daily life of older adults. *Geropsychiatry, 30*, 119–129. https://doi.org/10.1024/1662-9647/a000172

Hamaker, E. L. (2012). Why researchers should think within-person. In M. R. Mehl & T. S. Conner (Eds.), *Handbook of research methods for studying daily life* (pp. 43–61). New York, NY/ London, UK: The Guilford Press.

Hamaker, E. L., Asparouhov, T., Brose, A., Schmiedek, F., & Muthén, B. O. (2017). At the frontiers of modeling intensive longitudinal data: Dynamic structural equation models for the affective measurements from the COGITO study. *Multivariate Behavioral Research, 53*(6), 820–841.

Hay, E. L., & Diehl, M. (2011). Emotion complexity and emotion regulation across adulthood. *European Journal of Ageing, 8*, 157–168. https://doi.org/10.1007/s10433-011-0191-7

Heckhausen, J. (1995). A life-span theory of control. *Psychological Review, 102*(2), 284–304. https://doi.org/10.1037/0033-295X.102.2.284

Heckhausen, J., Wrosch, C., & Schulz, R. (2010). A motivational theory of life-span development. *Psychological Review, 117*(1), 1–53. https://doi.org/10.1037/a0017668.A

Hoppmann, C. A., & Riediger, M. (2009). Ambulatory assessment in lifespan psychology: An overview of current status and new trends. *European Psychologist, 14*, 98–108. https://doi.org/10.1027/1016-9040.14.2.98

John, O. P., Naumann, L. P., & Soto, C. J. (2008). Paradigm shift to the integrative big Five Trait Taxonomy: History, measurement, and conceptual issues. In O. P. John, R. W. Robins, & L. A. Pervin (Eds.), *Handbook of personality: Theory and research* (pp. 114–158). New York, NY: Guilford Press.

Kanning, M., Ebner-Priemer, U. W., & Brand, R. (2012). Autonomous regulation mode moderates the effect of actual physical activity on affective States: An ambulant assessment approach to the role of self-determination. *Journal of Sport & Exercise Psychology, 34*(2), 260–269.

Kanning, M., Ebner-Priemer, U. W., & Schlicht, W. M. (2015). Using activity triggered e-diaries to reveal the associations between physical activity and affective states in older adult's daily living. *International Journal of Behavioral Nutrition and Physical Activity, 12*(1), 111. https://doi.org/10.1186/s12966-015-0272-7

Kanning, M., & Hansen, S. (2017). Need satisfaction moderates the association between physical activity and affective states in adults Aged 50+: an activity-triggered ambulatory assessment. *Annals of Behavioral Medicine, 51*, 18–29. https://doi.org/10.1007/s12160-016-9824-6

Kashdan, T. B., & Rottenberg, J. (2010). Psychological flexibility as a fundamental aspect of health. *Clinical Psychology Review, 30*, 865–878. https://doi.org/10.1016/j.cpr.2010.03.001

Kunzmann, U., Kappes, C., & Wrosch, C. (2014). Emotional aging: A discrete emotions perspective. *Frontiers in Psychology, 5*. https://doi.org/10.3389/fpsyg.2014.00380

Kuppens, P., Oravecz, Z., & Tuerlinckx, F. (2010). Feelings change: Accounting for individual differences in the temporal dynamics of affect. *Journal of Personality and Social Psychology, 99*, 1042–1060. https://doi.org/10.1037/a0020962

Labouvie-Vief, G. (2003). Dynamic integration: Affect, cognition and the self in adulthood. *Current Directions in Psychological Science, 12*, 201–206. https://doi.org/10.1046/j.0963-7214.2003.01262.x

Larsen, J. T. (2017). Holes in the case for mixed emotions. *Emotion Review, 9*, 118–123. https://doi.org/10.1177/1754073916639662

Mroczek, D. K., & Almeida, D. M. (2004). The effect of daily stress, personality, and age on daily negative affect. *Journal of Personality, 72*, 355–378. https://doi.org/10.1111/j.0022-3506.2004.00265.x

Neupert, S. D., Almeida, D. M., & Charles, S. T. (2007). Age differences in reactivity to daily stressors: The role of personal control. *Journal of Gerontology: Psychological Sciences, 62B*(4), 216–225. Retrieved from https://oup.silverchair-cdn.com/oup/backfile/Content_public/Journal/psychsocgerontology/62/4/10.1093/geronb/62.4.P216/2/P216.pdf?Expires=1502279827&Signature=LrF6l83bTXq0FNvSAaFuH72Z~VEIaUIuWWjNpWW-tphQ89VedHofhvDxNzTZdV4Vf-F1JWJE6PCcD9lO7Q~bR0XFc9uHsE~5r

Ong, A. D., & Bergeman, C. S. (2004). The complexity of emotions in later life. *Journal of Gerontology: Psychological Sciences, 59*, 117–122. https://doi.org/10.1093/geronb/59.3.P117

Pinquart, M. (2001). Correlates of subjective health in older adults: A meta-analysis. *Psychology and Aging, 16*(3), 414–426. https://doi.org/10.1037//0882-7974.16.3.414

Ram, N., & Gerstorf, D. (2009). Time-structured and net intraindividual variability: Tools for examining the development of dynamic characteristics and processes. *Psychology and Aging, 24*, 778–791. https://doi.org/10.1037/a0017915

Ready, R. E., Åkerstedt, A. M., & Mroczek, D. K. (2012). Emotional complexity and emotional well-being in older adults: Risks of high neuroticism. *Aging & Mental Health, 16*, 17–26. https://doi.org/10.1080/13607863.2011.602961

Ready, R. E., & Robinson, M. D. (2008). Do older individuals adapt to their traits?: Personality-emotion relations among younger and older adults. *Journal of Research in Personality, 42*, 1020–1030. https://doi.org/10.1016/j.jrp.2008.02.004

Roberts, B. W. (2009). Back to the future: Personality and assessment and personality development. *Journal of Research in Personality, 43*, 137–145. https://doi.org/10.1016/j.jrp.2008.12.015

Röcke, C., & Brose, A. (2013). Intraindividual variability and stability of affect and well-being. *Geropsychiatry, 26*, 185–199. https://doi.org/10.1024/1662-9647/a000094

Röcke, C., Brose, A., & Kuppens, P. (2018). Emotion dynamics in older age. In P. M. Cole & T. Hollenstein (Eds.), *Emotion regulation: A matter of time* (pp. 179–207). New York, NY: Routledge.

Röcke, C., Hoppmann, C. A., & Klumb, P. L. (2011). Correspondence between retrospective and momentary ratings of positive and negative affect in old age: Findings from a one-year measurement burst. *Journal of Gerontology: Psychological Sciences, 66B*, 411–415. https://doi.org/10.1093/geronb/gbr024

Röcke, C., Li, S., & Smith, J. (2009). Intraindividual variability in positive and negative affect over 45 days: Do older adults fluctuate less than young adults? *Psychology and Aging, 24*, 863–878. https://doi.org/10.1037/a0016276

Rosenberg, E. L. (1998). Levels of analysis and the organization of affect. *Review of General Psychology, 2*, 247–270. https://doi.org/10.1037/1089-2680.2.3.247

Russell, J. A. (2017). Mixed emotions viewed from the psychological constructionist perspective. *Emotion Review, 9*(2), 111–117. https://doi.org/10.1177/1754073916639658

Scheibe, S., & Carstensen, L. L. (2010). Emotional aging: Recent findings and future trends. *Journal of Gerontology: Psychological Sciences, 65B*, 135–144. https://doi.org/10.1093/geronb/gbp132

Schulz, R., & Heckhausen, J. (1996). A life span model of successful aging. *American Psychologist, 51*(7), 702–714. Retrieved from https://pdfs.semanticscholar.org/d87c/3eefb33a833668d7d97859e96d1f7d882d4f.pdf

Schwarzer, R., & Leppin, A. (1991). Social support and health: A theoretical and empirical overview. *Journal of Social and Personal Relationships, 8*, 99–127.

Schwarzer, R., & Luszczynska, A. (2008). How to overcome health-compromising behaviors the health action process approach. *European Psychologist, 13*(2), 141–151. https://doi.org/10.1027/1016-9040.13.2.141

Scott, S. B., Sliwinski, M. J., & Blanchard-Fields, F. (2013). Age differences in emotional responses to daily stress: The role of timing, severity, and global perceived stress. *Psychology and Aging, 28*, 1076–1087. https://doi.org/10.1037/a0034000

Steel, P., Schmidt, J., & Shultz, J. (2008). Refining the relationship between personality and subjective well-being. *Psychological Bulletin, 134*, 138–161. https://doi.org/10.1037/0033-2909.134.1.138

Trull, T. J., & Ebner-Priemer, U. W. (2013). Ambulatory assessment. *Annual Review of Clinical Psychology, 9*, 151–176. https://doi.org/10.1146/annurev-clinpsy-050212-185510

Watson, D., Clark, L. A., & Tellegen, A. (1988). Development and validation of brief measures of positive and negative affect: The PANAS scales. *Journal of Personality and Social Psychology, 54*, 1063–1070. https://doi.org/10.1037/0022-3514.54

World Health Organization. (2015). *World report on aging and health*. Retrieved from http://apps.who.int/iris/bitstream/10665/186463/1/9789240694811_eng.pdf?ua=1

Wrzus, C., Luong, G., Wagner, G. G., & Riediger, M. (2015). Can't get it out of my head: Age differences in affective responsiveness vary with preoccupation and elapsed time after daily hassles. *Emotion, 15*, 257–269. https://doi.org/10.1037/emo0000019

Wrzus, C., Müller, V., Wagner, G. G., Lindenberger, U., & Riediger, M. (2013). Affective and cardiovascular responding to unpleasant events from adolescence to old age: Complexity of events matters. *Developmental Psychology, 49*(2), 384–397. https://doi.org/10.1037/a0028325

Chapter 6
Coordinated Data Analysis: A New Method for the Study of Personality and Health

Sara J. Weston, Eileen K. Graham, and Andrea M. Piccinin

The proliferation of panel studies and other publicly-available datasets has greatly advanced the study of healthy aging, as these datasets serve as a low-burden avenue for examining longitudinal relationships, often including real-world outcomes. However, as the Open Science Reform Movement has brought attention to common pitfalls in data analysis (Munafò et al., 2017), it is clear that careful steps must be taken to ensure robust and rigorous analyses of panel study data. The current chapter will review the benefits and pitfalls of panel studies and then discuss one method of rigorous data analysis: the coordinated analysis.

6.1 The Benefits of Panel Studies

The nature of psycho-social change being one of the primary interests of developmental and aging researchers (Baltes & Nesselroade, 1979), panel studies and other similar publicly-available datasets are widely used in psychology. Panel studies – the repeated assessment of large samples across years or decades – allow for estimations of change in psycho-social constructs across developmental periods and after major life events (Vartanian, 2010). More than just being longitudinal in nature, panel studies lend several advantages to a research project.

S. J. Weston (✉)
Department of Psychology, University of Oregon, Eugene, OR, USA
e-mail: sweston2@uoregon.edu

E. K. Graham
Department of Medical Social Sciences, Northwestern University, Chicago, IL, USA

A. M. Piccinin
Department of Psychology, University of Victoria, Victoria, BC, Canada

P. L. Hill, M. Allemand (eds.), *Personality and Healthy Aging in Adulthood*, International Perspectives on Aging 26,
https://doi.org/10.1007/978-3-030-32053-9_6

First, many are large in size (i.e., assess many individuals), allowing for the estimation of small but meaningful effect sizes. For example, the Midlife in the United States Study (MIDUS, Brim et al., 1999) has been used to link sense of purpose to net worth (Hill, Turiano, Mroczek, & Burrow, 2016). The effect size estimated was small (with a standardized regression coefficient of about 0.04) but translates into a large real-world effect (for every standard deviation increase in purpose, participants had on average $20K greater net worth). Large samples also allow for better group comparisons. The relationship of personality to health, for instance, can differ by ethnicity; large samples, like in the Hawaii Personality and Health Cohort (Hampson & Goldberg, 2006), allow for the estimation of these marginal effects without undue loss of statistical power (McClelland & Judd, 1993).

As a second advantage, panel studies are often well-funded and well-staffed, allowing for the collection of data that would otherwise be too expensive or time-consuming to collect. The UK Biobank (Sudlow et al., 2015), for instance, has genetic data on more than half a million participants to date. The Health and Retirement Study has biomarker and prescription drug use data on thousands of its participants. Individual research labs, even teams of labs, would struggle to find the money and manpower to assess genetics or biomarkers in samples of this magnitude; to then link such data to decades worth of survey research is almost unthinkable.

Which brings us to the third advantage of panel studies: time and money. Accessing a panel study is often free or, if not, less costly than paying for the cost of interviewing and assessing thousands of participants. This is especially useful for researchers with limited financial resources, such as early career researchers – students, post-docs, untenured faculty – and for researchers who work in teaching-focused environments without grant funding. Moreover, developmental and lifespan questions often require years, even decades, of data collection, making such research difficult for anyone. Moreover, many research questions about the lifespan will need a lifespan's worth of data.

6.2 Challenges in Analyzing Panel Studies

Analysis of panel studies and other publicly available datasets is not without pitfalls. We have named several panel studies already in this chapter (e.g., the UK Health and Lifestyle Survey, the MIDUS, the UK Biobank). While each has been used to address similar questions, the published results are not directly comparable. For example, each has been used to assess the relationship of neuroticism to health. Neuroticism is related to greater mortality in the UK Health and Lifestyle Survey (Shipley, Weiss, Der, Taylor, & Deary, 2007). In the MIDUS, neuroticism was unrelated to mortality risk (Turiano et al., 2015). Finally, using the UK Biobank, Gale et al. (2017) found that neuroticism was associated with a reduction in mortality risk. But these results are not directly comparable. Each panel study used a different measure of neuroticism – the first and third used the Eysenck Personality Inventory with a binary response choice, while the second used the MIDI with a 4-point Likert scale. Each study used different covariates: while all controlled for age and gender, the second also controlled for race, marital status and education; the third study included at least a dozen additional covariates. The time between personality assessment and mortality varied across studies: 21 years, 11 years, and 14 years, respectively. And each study used their own approach to variable transformation and standardization. There are other differences of course, including the location of the study, ages of the participants at baseline, year of data collection, etc. It is clear to see why comparing these findings cannot clarify the link between neuroticism and mortality. There are too many differences between the samples and analyses to make sense of why any differences in results emerge.

Adjusting for these differences is difficult under the best of circumstances; however, the scale of panel studies allows for model overfitting, which can bias published results. By overfitting, we mean mistakenly fitting a model to sample-specific noise (for a discussion, see Yarkoni & Westfall, 2017). Overfit models are problematic as they can include spurious relationships that will not generalize outside of the sample at hand. One of the easiest ways to overfit a model is to include too many terms, such as too many covariates or too many polynomials and interactions. Panel studies can include thousands of variables and tens of thousands of subjects; motivated researchers have the opportunity to fit many different models, on different subsets of the data. These models may ultimately be over-complex, over-fitted and p-hacked. P-hacking, or modifying analyses with the goal of obtaining a p-value smaller than the significance threshold (Simmons, Nelson, & Simonsohn, 2011), is often assumed to be an intentional act of the researcher, but even honest researchers can unknowingly engage in this practice. For example, researchers can bias results if they have previous knowledge of a dataset; given that researchers often return to the same panel studies multiple times, this is not only possible but likely (see Weston, Ritchie, Rohrer, & Przybylski, 2019). Even without deliberate p-hacking, prior knowledge of datasets precludes the traditional use of p-values, which can only be interpreted under the assumption that the statistical test is independent of any other statistical test; prior knowledge violates this assumption.

Unfortunately, our modern system of publishing research relies heavily on the use of statistical significance. Therefore, researchers who wish to publish secondary data analysis must still use statistical significance, despite the violation of the independent test assumption. Responsible researchers will seek additional ways to improve the robustness of such tests. Luckily, there is a proliferation of statistical tools for such tasks. Machine learning techniques can be used to evaluate the robustness of effect sizes (Yarkoni & Westfall, 2017). The use of preregistration and other open science practices can effectively limit researcher degrees of freedom and enhance transparency (Weston, Ritchie, et al., 2019). Researchers who use publicly available panel studies have a particularly useful tool: more panel studies. In other words, researchers can leverage the availability of large datasets to answer simple but important research questions, limit researcher degrees of freedom, and test moderators imposed by time, location and study design. This process is referred to as coordinated analysis.

6.3 Coordinated Analysis

A coordinated analysis is the analysis of multiple independent datasets in a way that optimizes the comparison of results (Hofer & Piccinin, 2009). The primary goal of a coordinated analysis is to maximize the opportunities for direct comparisons of results. To achieve this goal, researchers analyze identical (or nearly identical) statistical models across multiple datasets but preserve other differences in data (e.g., measurement instrument, cohort). These meaningful differences allow researchers to estimate the generalizability of an effect across time, culture and instrument or, in the case where differences emerge, identify potential moderators of an effect. Often, the results from individual datasets are compared using the tools of meta-analysis, which allow for statistical estimating of heterogeneity and the moderating effect of sample-level characteristics, such as location or year of data collection.

Coordinated analysis is a subset of a larger group of analytic tools, referred to as integrative data analysis. A coordinated analysis can be collaborative if using proprietary datasets or if seeking to involve many researchers; it may also be an independent exercise if a researcher limits himself or herself to only directly accessible datasets.

6.4 Recommendations for Coordinated Analysis

The steps of coordinated data analysis follow much the same process as the scientific method: a research question is answered by testing specific hypotheses with data. This process quickly becomes complex, as these steps are simultaneously implemented in many different datasets and then integrated into a concise, if not simple, picture. Consequently, researchers should be careful to limit the complexity of their hypotheses and statistical models. Moreover, coordinated analyses often become team efforts, especially if using proprietary data. The authors of this chapter have each participated in coordinated analyses conducted by over two dozen researchers at a time. While the logistics of coordinating analyses across many datasets (and in some cases, several continents) is made easier by the Internet, we note that ensuring all researchers are on the same page is no simple task. It is with this in mind that we make the following recommendations:

1. *Study Simple Yet Foundational Relationships*

Simple models are the key to successful coordinated analyses, for two reasons. First, overly complex models are difficult to integrate across multiple datasets. For example, structural equation models often require modification to converge and adequately fit one dataset; attempting to fit an identical model across multiple datasets with adequate fit may be a Sisyphean task. Second, overly complex models limit the number of datasets eligible for inclusion in a coordinated analysis, as these models include more variables, sometimes requiring multiple measurement occasions with specific temporal ordering. Inclusion of an extra covariate in a linear model can reduce the datasets in which the full model can be estimated. We recommend that researchers limit their hypotheses to variables deemed essential and to answer questions using the simplest form of the model possible.

When developing a research question, researchers should consider that coordinated analysis is most useful for establishing foundational relationships or processes and estimating the generalizability of those key effects. For example, personality psychologists are interested in the relationship between trait conscientiousness and mortality; cognitive psychologists are interested in the growth and decline in working memory over time. Findings such as these are central to forming theories and studying complex processes. Therefore, in order to best develop new theories, psychologists must establish the existence of such relationships, the size of the relationship, and the conditions under which the association is strongest. Coordinated analysis provides an efficient way to estimate all three of these key characteristics. We note that, while this method may be used for exploratory research as well, it is not well-suited for such analyses given the time and planning required; as we will discuss below, coordinated analyses often include extensive coordination with multiple team members, many analytic decisions prior to analysis, and the development and testing of analytic scripts prior to data analysis. These processes are facilitated by clear research aims and specific, confirmatory hypotheses. In other words, the labor involved in a coordinated analysis can make it difficult to fully explore a series of datasets, though it can facilitate the generation of new hypotheses for current or subsequent projects (Hofer & Piccinin, 2009, 2010; Piccinin et al., 2013).

2. *Establish Minimum Inclusion Criteria*

Once the analytic model is specified, it becomes necessary to identify the criteria a particular dataset must meet for the model to be tested within that dataset and to be included in a coordinated analysis. Inclusion criteria may include the following: constructs measured, measurement instrument and quality, number and spacing of measurement occasions, populations sampled, and data quality checks. Ideally, simple hypotheses and simple models require fewer inclusion criteria. For example, a simple correlation test requires that a dataset measure only two constructs. It is important to consider the conditions under which a dataset is included or excluded from a coordinated analysis before beginning the task of gathering and organizing data. Doing so not only saves time, but reduces the risk of post-hoc changes.

The choice of inclusion criteria ultimately dictates the scope of the coordinated analysis, the ability to compare results, and the estimation of generalizability of findings. More stringent criteria – e.g., all datasets must use the same instrument to measure a construct of interest – allows for easier comparison of results across samples, but limits the generalizability of results to only that instrument. On the other hand, criteria regarding the quality of data, such as internal reliability cutoffs or necessary base rates for binary variables, reduces the noise in estimates. We recommend that researchers limit variable inclusion criteria to only those necessary to adequately fit the model. Doing so ensures the quality of results while allowing for the maximum number of datasets. For example, Graham, Weston, Gerstorf, et al. (2019) examined trajectories of Big Five personality traits using multi-level growth models in a coordinate analysis framework. The minimum requirement for a dataset to be included was three measurement occasions of at least one of the Big Five personality traits. Studies with a single trait were still included in the project and simply excluded from any analysis for which they did not have the appropriate data.

Once minimum inclusion criteria are defined, researchers must comprehensively search for available datasets. There are a number of excellent resources available for identifying appropriate datasets for a given project. Many data repositories have archived publicly available data so that anyone can register and access them. For example, ICPSR houses data on social research (broadly speaking) and contains a number of specialized "collections" of datasets (e.g., the Resource Center for Minority Data [RCMD], and the National Archive of Computerized Data on Aging [NACDA]). The National Institute for Mental Health (NIMH) has a similarly organized repository, containing specialized collections of archived mental health data, sorted by subject area (e.g., the Osteoarthritis Initiative, the National Database for Autism Research). There are also a number of international data networks, such as Ageing Trajectories of Health: Longitudinal Opportunities and Synergies (ATHLOS). Many of the datasets in the above-mentioned repositories are publicly available.

However, limiting oneself only to public data could introduce new bias into the results of a project. Therefore, we recommend also exploring access to non-public data as well. The Integrative Analysis of Longitudinal Studies of Aging and Dementia (IALSA) is one such network. The IALSA is a network of studies of aging from around the world, many of which are not publicly accessible for legal or ethical reasons. The network has created a pipeline by which individual researchers can use these data for their research purposes while not accessing the data itself. This is achieved through the use of developing and sharing analytic scripts. An individual researcher who is interested in conducting a coordinated analysis will write an analytic script (for example, an R script) that details precisely which analyses will be run; with what options, assumptions, or conditions; and in what order. The analytic script may also include essential transformations of variables, creations of indices, and post-hoc follow-up tests or data-visualization. Importantly, these scripts are written to work on any dataset with the appropriate variables (more information on scripting is including in the next section). This script is shared with a data-analyst associated with the dataset of interest. The analyst runs the analysis script on their data and sends the resulting output back to the individual researcher who can incorporate the output into a meta-analysis and write-up of results. The IALSA website, currently powered by Maelstrom, contains a variable search tool that can be used to identify studies with the variables/constructs needed. Contact information for each study is provided if an IALSA-affiliated study described in the Maelstrom catalogue is a good fit for a given project. The IALSA network and its website should be cited/acknowledged in any product arising from knowledge gained from the site.

3. *Leverage Scripting Technology*

While scripting languages are not new to statistical analysis, they have increased in appeal during the renewed focus on reproducibility in the social sciences. As a consequence, it is easier to not only write analytic scripts but share them through online platforms (e.g., the Open Science Framework or GitHub) and collaboratively develop them. We therefore recommend that researchers conducting coordinated analyses rely heavily on scripting languages. The benefits of using this method include (1) ensuring that models are specified identically across datasets, (2) the capacity to immediately summarize and integrate results, and (3) the integration of results directly into the manuscript. Regarding ensuring identical specification, coordinated analyses are meant to examine the heterogeneity of a particular relationship across multiple samples, measurement instruments, and/or time points. Therefore, other sources of variability should be avoided at all costs. One potential source of variability is the use of different analyses across datasets. For a solo researcher conducting all analyses, this means care should be taken to ensure that the models estimated on each dataset are identical, including decision rules surrounding the identification and removal of outliers, the use of different underlying estimation procedures, when and how to transform variables, the process for dealing with missing data, and the modeling of dependencies between family members, to name a few. While the most conscientious researchers among us will clearly document every single analytic decision and check that each decision was applied consistently across datasets, there is still room for human error if the analyst repeats every step by hand. For a team of data analysts, the potential for discrepancies increases exponentially, as data analysts may vary in their expertise or experience with the analysis at hand, their opinions and standard operating procedures for the analysis, and even the statistical software they use to estimate models, given that different softwares often have different default settings. For example, the default ANOVA sums of squares calculation for SPSS is Type III, but for R is Type I. Without scripting technology, coordination could only be achieved by detailed and specific communication of analytic decisions among these analysts and the appropriate implementation of these decisions by each analyst.

However, if each dataset is analyzed using an identical script, then consistency across analyses is more likely, regardless of the carefulness of a researcher or the communication skills of a team. An effective analytic script employs logical statements to ensure that a dataset meets the minimum inclusion criteria, checks that data are properly cleaned and coded, identifies which variables are available and thus which models can be tested, estimates the appropriate models, extracts descriptive and inferential statistics, and saves the information necessary for writing the final manuscript and estimating overall (meta-analytic) effects, if those are to be included. In other words, an effective analytic script can be run by any data analyst on any dataset without requiring idiosyncratic changes. This of course requires additional effort upfront. Scripts should be tested on simulated data to ensure they work properly. But this effort ultimately pays off when it comes time to analyze many datasets simultaneously. Not only does this step in a coordinated analysis go quickly, but there are fewer opportunities for human error. An identical script ensures identical analyses across studies.

Scripting technology also facilitates the summary and integration of results across datasets. Analytic scripts can be used to identify, extract, and name key statistics in a systematic fashion, allowing for those key statistics to be efficiently integrated into a summary. For example, imagine conducting a coordinated analysis on the correlation between two variables. The analytic script used on each dataset could calculate and generate a name that identifies the statistic and the dataset, e.g., *r_A, r_B, r_C* for datasets A, B, and C. It would then be easy to identify all statistics that include the prefix "*r_*" and combine these into an average correlation. In fact, such scripts can be written prior to analyses conducted on individual datasets, meaning a researcher conducting a coordinated analysis can easily re-estimate her meta-analytic effects as new data come in or after mistakes are identified and fixed.

Finally, there have been recent advances in integrating open source statistical software, like R, with open source text formatting software, like LaTex. In other words, it is becoming easier and easier to automatically populate the text of a manuscript with statistics generated through R and to incorporate and even reference summary tables and figures. These integrations both increase the efficiency with which researchers can summarize and communicate results and also reduce human error in the generation of figures and tables.

4. *Preregister Analyses Prior to Data Cleaning*

Preregistration, the act of describing in detail the analysis plan for a study prior to working with data, benefits a project on many fronts. These benefits are not limited to coordinated analysis, but apply to all research. Preregistration was developed to increase transparency and reduce researcher degrees of freedom. By making analytic decisions prior to viewing results, researchers can avoid making data-dependent decisions, which violate assumptions of null hypothesis significance testing and invalidate the traditional interpretations of results (Gelman & Loken, 2013). Simply put, if a researcher sees a relationship in a dataset and then statistically tests that relationship, the test is no longer a test; this is analogous to flipping the coin and then calling heads or tails.

Similarly, preregistration can be used to designate which parts of a project are confirmatory tests and which are exploratory. Preregistration does not preclude the inclusion of additional analyses; rather, a preregistration makes clear which analyses were planned prior to data analysis and which were developed after having seen data. This distinction is crucial, as analytic choices informed by data violate the assumption of independence that is central to probability theory and hypothesis testing. For this reason, preregistration provides a tool for responding to reviewers who ask for more complex analyses in order to "find" a result they are satisfied with. Specifically, if researchers believe they must run additional analyses to be published, they can include those analyses to satisfy reviewers, but use the preregistration to justify caution in interpreting results and labeling them as exploratory. Publicly posting preregistrations with date-stamps counters the file-drawer problem, as a record of the study is available, even if the results are not published.

In our own work, we have found that preregistration is especially useful for coordinated analyses. The act of completing a preregistration, especially one that is detailed and thorough, helps to organize the process going forward. For example, preregistration forms may ask a researcher to outline a plan to enact when outliers are detected; this should then prompt a researcher to write syntax that tests for outliers and carry out this plan when they are found. Overall, anticipating not only the primary model, but the conditions that must be met to run this model and the contingency plans when the model fails aids researchers in developing working syntax and limiting data-dependent decisions. In addition, given that coordinated analyses are often conducted by teams of researchers, a preregistration can help ensure all team members are on the same page concerning the analyses and interpretations. Writing a preregistration and giving all team members an opportunity to read, refine, and agree to the preregistration limits conflict at later stages in the project.

We feel strongly that robust and replicable research in lifespan development need to approach the analysis planning stage using theory-driven analysis decisions, not data-driven analysis decisions. The latter can easily lead to model-overfitting, which by extension leads to non-replicable results. Some longitudinal modeling techniques by definition require data-driven decisions: for example, in growth mixture modeling, the first step in the model is to identify a set of groups in a dataset, and the second set using those groups to predict an outcome. The results of these models tend to be highly specific to a given dataset and are often impossible to replicate. One solution is to use cross-validation techniques, where each dataset in the coordinated analysis is subset into a "training" sample and a "test" sample, where the data-driven portion of the analysis decision are fine-tuned and coordinated across studies within the training samples using cross-validation techniques, then ultimately evaluated based on performance in the hold-out test samples.

5. *Consider Generalizability*

One of the strengths of coordinated analysis is the opportunity to formally test the generalizability of an effect across time, culture, population, and method. Establishing generalizability, or the limits thereof, is a key goal of research, as it can dramatically affect how scientists choose to study a phenomenon and the way they interpret results. If conscientiousness is associated with health only when researchers measure personality using a specific instrument, then we are less confident in this result. If extraversion follows the same developmental trajectory in all cultures, we are more confident. If cognitive ability predicts income in all cohorts except one, we look to historical events that may have interfered with this relationship. In short, coordinated analysis not only provides a high-powered and precise estimate of specific parameter; it can also test the degree to which that parameter estimate applies to the world population.

The tests of generalizability are only as good as the data. We therefore encourage researchers (1) seek datasets that collectively cover a range of cultures, timepoints, measurement instruments, or any variable which may affect the hypothesized relationship, and (2) examine as many moderating contexts as possible. The latter largely involves examining differences between different datasets: geographic location, cohort, measurement spacing, and instrument are all possible moderators of an effect. We recommend consulting Hofer and Piccinin (2009) for a discussion of common sources of between-study heterogeneity.

Estimating the degree of between-study heterogeneity is often a goal of meta-analysis, and so several metrics exist to serve this function. The I2 statistic (Higgins & Thompson, 2002) represents the proportion of variance in the estimates that is due to heterogeneity between the studies. While this statistic is commonly used, it is also subject to bias, as it assumes homogeneity of the within-study variances (which is almost never the case); other statistics, such as Q, estimate this heterogeneity with less bias (Crippa, Khudyakov, Wang, Orsini, & Spiegelman, 2016). These statistics are invariably accompanied by tests of significance, which may be used as a tool to avoid imprecise guessing or eyeballing of heterogeneity. We do not recommend one estimate over another but rather suggest researchers estimate several statistics and come to a more holistic conclusion. We also recommend estimating whether effect sizes can be predicted by study-level characteristics, such as year of study. We also note that consistency and generalizability can be tested within-study as well, if multiple measures of a construct are available, or if participants differ in age, gender, or other key characteristics. Include planned tests of heterogeneity between studies and tests of moderation within studies in a preregistration.

6.5 Coordinated Analysis in Action

In the following section, we illustrate the approach of coordinated analysis by detailing one specific example of its use. We start by summarizing the motivation of the study, and then describe the application of our five recommendations.

6.5.1 Motivation

The research project we describe here attempts to answer the following research question: Does the relationship between neuroticism and health depend on levels of conscientiousness? The relationship of neuroticism to health is less defined than many personality researchers may think. While typically neuroticism is associated with worse health – including greater engagement in unhealthy behaviors, increased risk of developing chronic conditions, and reduced lifespan – there are enough published null and even positive associations between neuroticism and health to conclude that the relationship between neuroticism and health may be more complex. This heterogeneity was first described by Friedman (2000) who suggested that such diverse findings could result from individuals high in neuroticism falling into two categories: the traditional neurotics, who deal with stress through maladaptive coping behaviors like smoking and drinking, and the healthy neurotics, who proactively address changes in their health and, consequently, detect major health concerns at earlier and more treatable stages. Implicit in this theory is the presence of a moderator, or a third variable which determines whether a person high in neuroticism is healthy or unhealthy.

While the specific moderator of neuroticism was not identified through this theory – and indeed, several have been examined empirically, including SES (Elliot, Turiano, & Chapman, 2016; Hagger-Johnson et al., 2012; Weston, Hill, Edmonds, Mroczek, & Hampson, 2018), physical health (Weston et al., 2018), and even self-rated health (Gale et al., 2017) – the most promising candidate moderator is trait conscientiousness. Conscientiousness, the tendency to be rule-following, industrious, and orderly, is commonly presented as the trait needed to focus the natural worries of a person high in neuroticism on controllable domains (like health) and address problems in this domain through adaptive and healthy behaviors. Often individuals high in both traits are described as likely being "vigilant" towards their health (Turiano, Mroczek, Moynihan, & Chapman, 2013, p. 87; Weston & Jackson, 2015, p. 7).

Since the introduction of the healthy neuroticism concept almost two decades ago, the idea is frequently found in theoretical work describing the relationship of personality to health (e.g., Friedman, 2019; Murray & Booth, 2015). Despite the popularity of this theory, the empirical evidence supporting this interaction is limited and mainly focused on smoking and alcohol consumption (cf. Turiano, Whiteman, Hampson, Roberts, & Mroczek, 2012; Vollrath & Torgersen, 2002). Sometimes, the interaction is only present in three-way interactions, such as the interaction of conscientiousness, neuroticism, and gender to predict mortality risk in Friedman, Kern, and Reynolds (2010), or in subsamples, such as only among individuals diagnosed with some chronic conditions in Weston and Jackson (2015). Moreover, a number of empirical studies show no relationship of this interaction to health behaviors (e.g., Atherton, Robins, Rentfrow, & Lamb, 2014), adaptive responding to health news (Weston & Jackson, 2016), or more distal outcomes, such as frailty (Stephan, Sutin, Canada, & Terracciano, 2017). It is unclear whether these discrepant results are due to each of these analyses being underpowered to detect the interaction (McClelland & Judd, 1993), or if this interaction is only related to health among some populations or when measured using certain instruments, or if the true interaction effect is null and the published significant results are simply capturing spurious effects. What is needed to determine the true size of this interaction effect is a highly powered study that can examine meaningful differences between samples, namely, a coordinated analysis. We will now describe how we either followed the recommendations of this chapter to develop our coordinated analysis project or, when we did not follow our own advice, how this affected our project.

6.5.2 *Choosing Simple Yet Foundational Relationships*

When designing the coordinated analysis best suited to test whether the relationship of neuroticism to health was indeed moderated by conscientiousness, we first attempted to define the simplest relationship underlying this theory. This proved to be a difficult task, specifically because the empirical evidence that inspired the original description of healthy neuroticism (Friedman, 2000) was entirely based on discrepancies in the neuroticism-mortality association. However, the bulk of the published empirical evidence for this interaction examined health behaviors, most often smoking and alcohol consumption (e.g., Turiano et al., 2012), as these are fairly straightforward to assess and the more proximal associates with personality.

Choosing one outcome over another would impact the interpretation of the healthy neuroticism theory: examining mortality would be a direct test of the primary theory, that is, does conscientiousness help us to reconcile major differences in the relationship between neuroticism and mortality, but would remain silent on the processes underlying these differences. Examining behaviors, on the other hand, would help to clarify how neuroticism may be adaptive, but not whether those behaviors would have a substantial impact on one's survival. We chose to examine both mortality and behavior, specifically smoking, alcohol consumption, and physical activity. We also examined chronic condition status (hypertension, diabetes, and heart condition) for good measure. These outcomes were relegated to three separate but parallel projects, a coordinated-analysis, if you will. Analytic decisions were repeated across the three projects, to the extent possible.

Ultimately, our analysis examines the interaction of neuroticism by conscientiousness on each outcome in question. We included several covariates that were known to be related to neuroticism and health or conscientiousness and health: age, reported gender, education level, and any other Big Five personality traits (agreeableness, openness, extraversion) measured concurrently with neuroticism and conscientiousness. Importantly, each model was tested with and without covariates, to examine sensitivity to the inclusion of these additional variables.

We note here that the choice of seven outcome variables is not in line with the recommendation to keep analyses simple. However, given that these models were nearly identical across outcome type, we were able to conduct the analyses efficiently and also provide a clear synopsis of the study. Moreover, in tension between simplicity and thoroughness, we decided it was better, here, to err on the side of thoroughness, as this would best illuminate the boundaries of the healthy neuroticism effect.

6.5.3 Establish Minimum Inclusion Criteria

We limited study selection to the IALSA network, which is sufficiently large to capture many populations at several time points with a variety of measurement instruments. Our inclusion criteria were broad: studies were included in the analyses if, during at least one wave of data collection, participants were measured on trait neuroticism, trait conscientiousness and at least one of the seven outcomes, in addition to age, gender, and education level. We identified 15 datasets that met these criteria, all of which had at least one health behavior and one chronic condition variable; 13 datasets included mortality information.

6.5.4 *Leverage Scripting Technology*

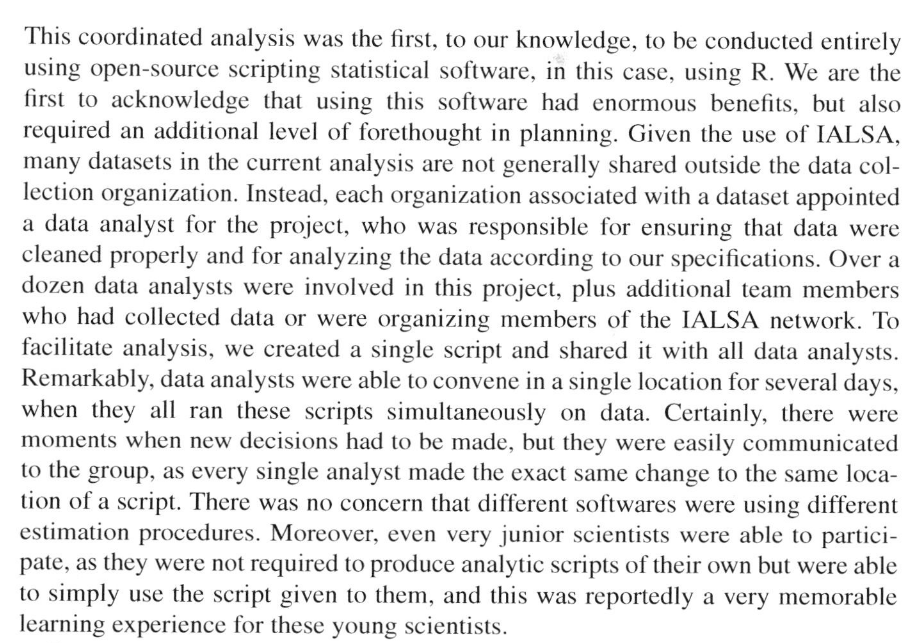

This coordinated analysis was the first, to our knowledge, to be conducted entirely using open-source scripting statistical software, in this case, using R. We are the first to acknowledge that using this software had enormous benefits, but also required an additional level of forethought in planning. Given the use of IALSA, many datasets in the current analysis are not generally shared outside the data collection organization. Instead, each organization associated with a dataset appointed a data analyst for the project, who was responsible for ensuring that data were cleaned properly and for analyzing the data according to our specifications. Over a dozen data analysts were involved in this project, plus additional team members who had collected data or were organizing members of the IALSA network. To facilitate analysis, we created a single script and shared it with all data analysts. Remarkably, data analysts were able to convene in a single location for several days, when they all ran these scripts simultaneously on data. Certainly, there were moments when new decisions had to be made, but they were easily communicated to the group, as every single analyst made the exact same change to the same location of a script. There was no concern that different softwares were using different estimation procedures. Moreover, even very junior scientists were able to participate, as they were not required to produce analytic scripts of their own but were able to simply use the script given to them, and this was reportedly a very memorable learning experience for these young scientists.

Because we used scripting technology, we were not only able to complete analyses on each individual dataset within a matter of hours, but it took only minutes to integrate these results into a series of overall weighted effects and display them using forest plots, or a graphical display of the estimates and confidence intervals of each individual dataset and the overall, weighted effect size and confidence interval (Lalkhen & McCluskey, 2008). The primary benefit of having written our integration scripts ahead of time was that it gave our team the opportunity to process results immediately and discuss the likelihood that this interaction sufficiently explained heterogeneity in the literature. Finally, we integrated our scripts into our manuscript using R and LaTex, which has allowed us to easily write up the results of our coordinated analysis and even make adjustments as analysts have identified mistakes in data cleaning.

6.5.5 *Preregister Analysis Prior to Data Cleaning*

For the healthy neuroticism project, we registered analyses on the Open Science Framework (osf.io/kcquy), which increased the confidence we have in interpreting our results. We anticipate that this preregistration will help both ensure the results are robust and also protect us from ad-hoc analyses proposed by potential reviewers. We have already benefited from the use of preregistration prior to submitting our manuscript for peer review. Specifically, during the data analysis meeting, it was proposed that age might further moderate the neuroticism by conscientiousness interaction. Regardless of the theoretical merits of this idea, it is ill-advised to design new analyses after seeing the results, as this violates the assumptions of null hypothesis significance testing (Weston, Ritchie, et al., 2019). Rather than simply throw out these analyses, we were able to add them to the project (efficiently, thanks to the use of scripting software) and present these analyses under the header of Unregistered Analyses. That is, we were not precluded from performing any analyses, allowing for the examination of any potentially important effect; however, we avoided problematic interpretations of these analyses by clearly labeling them as exploratory. (There was no evidence of this three-way interaction.)

6.5.6 Consider Generalizability

The choice of datasets informed our ability to estimate generalizability. One especially important factor – the measurement instrument used to assess neuroticism – was diversely sampled, as across these 15 datasets, we were able to compare results using five common personality scales: the NEO-FFI (Costa & McCrae, 1989), the NEO-PI-R (Costa & McCrae, 2008), the BFI (John, Donahue, & Kentle, 1991), the MIDI (Lachman & Weaver, 1997), and Goldberg et al. (2006)'s 50 Adjectives. In other ways, however, our generalizability was limited. All studies came from mainly white, industrialized (American, European or Australian) countries, the majority of participants were over the age of 55, and our measures were almost exclusively self-report (with the obvious exception of mortality).

In short, while a few analysts found statistically significant interactions of neuroticism and conscientiousness on some health behaviors and outcomes, overall, there was fairly consistent evidence that the interaction effect size did not differ from 0 (Turiano et al., 2019; Weston, Graham, et al., 2019). The primary exceptions to this was when conscientiousness was low, neuroticism was associated with more smoking and less activity; when conscientiousness is high, neuroticism was not associated with health behaviors (Graham, Weston, Turiano, et al., 2019). We found no evidence that the use of one particular scale changed either the relationship of neuroticism to health, the relationship of conscientiousness to health, or the interaction. The ability to examine the role of specific personality scales was an important contribution of our coordinated analysis. An important note about the results of these coordinated analysis: the interaction of conscientiousness and neuroticism predicting smoking and physical activity was only statistically significant in the overall meta-analysis and within only two or three datasets. A primary benefit of coordinated analysis is the opportunity to put these effect sizes into context: there was no heterogeneity in the size of the effects and, for both analyses, the effect sizes were quite small. Taken together with the significant interactions, we can propose that the moderation effect is present but quite small. We would have been less certain in this statement if we had run a single study and been unable to determine whether the effect size was unusual.

6.5.7 Summary of Healthy Neuroticism Coordinated Analysis

This is an example of when coordinated analysis is an appropriate and useful tool for research. The research question in this study – whether the interaction of neuroticism and conscientiousness associated with health – is both simple and yet inadequately answered by the published literature. We were able to assemble a relatively large number of panel studies to test this question using several different outcomes. We concluded that relatively few studies yielded a significant interaction, that the weighted average effect across all studies was null, and that this effect size was not further moderated by personality scale or age. Moreover, by using scripting technology, we were able to efficiently ensure that no additional variability was added to our study, quickly synthesize results, spark discussion among a large team of analysts and researchers, and foster a hands-on learning experience for junior researchers. Through preregistration, we were able to ensure the robust and careful interpretation of our results.

6.6 Conclusion

Coordinated analysis is one method of facilitating robust results using pre-existing data. Coordinated analysis involves the estimation of a simple relationship or model in multiple pre-existing datasets and then using these estimates to create an overall weighted effect size or model. With the improvement of open source scripting technology, coordinated analysis becomes easier, even across large teams of analysts. Scripting also allows for integration of results into manuscripts with minimal human error. Coordinated analysis can be used with other Open Science tools like preregistration to ensure the responsible interpretation of results. Finally, by paying attention to differences across datasets in location, time, measurement instruments and sample characteristics, the generalizability of results can be estimated as well.

The analysis of pre-existing data is difficult, especially as researchers return to the same datasets multiple times, and by becoming aware of relationships between study variables are unable to unambiguously interpret significance tests with one or even a couple datasets. Despite these issues, lifespan researchers must continue to use pre-existing datasets, as they are often the best tools for answering important lifespan research questions. The use of coordinated analysis can address the potential analytic problems that come with pre-existing data, by leveraging the availability of large datasets to answer simple but important research questions, limiting researcher degrees of freedom, and testing moderators imposed by time, location and study design.

References

Atherton, O. E., Robins, R. W., Rentfrow, P. J., & Lamb, M. E. (2014). Personality correlates of risky health outcomes: Findings from a large internet study. *Journal of Research in Personality, 50*, 56–60.

Baltes, P. B., & Nesselroade, J. R. (1979). *Longitudinal research in the study of behavior and development*. New York, NY: Academic.

Brim, O. G., Baltes, P. B., Bumpass, L. L., Cleary, P. D., Featherman, D. L., Hazzard, W. R., & Shweder, R. (1999). *Midlife in the United States (MIDUS 1), 1995–1996*. Ann Arbor, MI: Inter-university Consortium for Political and Social Research. https://doi.org/10.3886/ICPSR02760.v12.

Costa, P., & McCrae, R. (1989). *NEO five-factor inventory (NEO-FFI)*. Odessa, FL: Psychological Assessment Resources.

Costa, P. T., & McCrae, R. R. (2008). The revised NEO personality inventory (NEO-PI-R). In *The SAGE handbook of personality theory and assessment* (Vol. Vol. 2, pp. 179–198). Thousands Oaks, CA: SAGE.

Crippa, A., Khudyakov, P., Wang, M., Orsini, N., & Spiegelman, D. (2016). A new measure of between-studies heterogeneity in meta-analysis. *Statistics in Medicine, 35*, 3661–3675.

Elliot, A. J., Turiano, N. A., & Chapman, B. P. (2016). Socioeconomic status interacts with conscientiousness and neuroticism to predict circulating concentrations of inflammatory markers. *Annals of Behavioral Medicine, 51*, 240–250.

Friedman, H. S. (2000). Long-term relations of personality and health: Dynamisms, mechanisms, tropisms. *Journal of Personality, 68*, 1089–1107.

Friedman, H. S. (2019). Neuroticism and health as individuals age. *Personality Disorders, Theory, Research, and Treatment, 10*, 25–32.

Friedman, H. S., Kern, M. L., & Reynolds, C. A. (2010). Personality and health, subjective Well-being, and longevity. *Journal of Personality, 78*, 179–216.

Gale, C. R., Cukic, I., Batty, G. D., McIntosh, A. M., Weiss, A., & Deary, I. J. (2017). When is high neuroticism protective against death? Findings from UK biobank. *Psychological Science, 28*, 1345–1357.

Gelman, A., & Loken, E. (2013). *The garden of forking paths: Why multiple comparisons can be a problem, even when there is no "fishing expedition" or "p-hacking" and the research hypothesis was posited ahead of time*. New York, NY: Department of Statistics, Columbia University.

Goldberg, L. R., Johnson, J. A., Eber, H. W., Hogan, R., Ashton, M. C., Cloninger, C. R., & Gough, H. G. (2006). The international personality item pool and the future of public-domain personality measures. *Journal of Research in Personality, 40*, 84–96.

Graham, E. K., Weston, S. J., Gerstorf, D., Yoneda, T. B., Booth, T., Beam, C. R., Petkus, A. J., Rutsohn, J. P., Drewelies, J., Hall, A. N. et al. (2019). *Trajectories of Big Five personality traits: A coordinated analysis of 16 longitudinal studies*. Manuscript under review.

Graham, E. K., Weston, S. J., Turiano, N. A., Aschwanden, D., Booth, T., Harrison, F., James, B. D., Lewis, N. A., Makkar, S. R., Mueller, S. et al. (2019). *Is healthy neuroticism associated with health behaviors? A coordinated integrative data analysis*. Manuscript under review.

Hagger-Johnson, G., Roberts, B., Boniface, D., Sabia, S., Batty, G. D., Elbaz, A., … Deary, I. J. (2012). Neuroticism and cardiovascular disease mortality: Socioeconomic status modifies the risk in women (UK health and lifestyle survey). *Psychosomatic Medicine, 74*, 596–603.

Hampson, S. E., & Goldberg, L. R. (2006). A first large cohort study of personality trait stability over the 40 years between elementary school and midlife. *Journal of Personality and Social Psychology, 91*, 763–779.

Higgins, J. P., & Thompson, S. G. (2002). Quantifying heterogeneity in a meta-analysis. *Statistics in Medicine, 21*, 1539–1558.

Hill, P. L., Turiano, N. A., Mroczek, D. K., & Burrow, A. L. (2016). The value of a purposeful life: Sense of purpose predicts greater income and net worth. *Journal of Research in Personality, 65*, 38–42.

Hofer, S. M., & Piccinin, A. M. (2009). Integrative data analysis through coordination of measurement and analysis protocol across independent longitudinal studies. *Psychological Methods, 14*, 150–164.

Hofer, S. M., & Piccinin, A. M. (2010). Toward an integrative science of life-span development and aging. *Journals of Gerontology. Series B, Psychological Sciences and Social Sciences, 65*, 269–278.

John, O. P., Donahue, E. M., & Kentle, R. L. (1991). *The big five inventory—Versions 4a and 54*. Berkeley, CA: University of California–Berkeley, Institute of Personality and Social Research.

Lachman, M. E., & Weaver, S. L. (1997). *The midlife development inventory (MIDI) personality scales: Scale construction and scoring*. Waltham, MA: Brandeis University.

Lalkhen, A. G., & McCluskey, A. (2008). Statistics v: Introduction to clinical trials and systematic reviews. *Continuing Education in Anaesthesia, Critical Care & Pain, 8*, 143–146.

McClelland, G. H., & Judd, C. M. (1993). Statistical difficulties of detecting interactions and moderator effects. *Psychological Bulletin, 114*, 376–390.

Munafò, M. R., Nosek, B. A., Bishop, D. V., Button, K. S., Chambers, C. D., Du Sert, N. P., … Ioannidis, J. P. (2017). A manifesto for reproducible science. *Nature Human Behaviour, 1*, 0021.

Murray, A. L., & Booth, T. (2015). Personality and physical health. *Current Opinion in Psychology, 5*, 50–55.

Piccinin, A. M., & Hofer, S. M. (2008). Integrative analysis of longitudinal studies on aging: Collaborative research networks, meta-analysis, and optimizing future studies. In S. Hofer & D. Alwin (Eds.), *Handbook on cognitive aging: Interdisciplinary perspectives* (pp. 446–476). Thousand Oaks, CA: SAGE.

Piccinin, A. M., Muniz-Terrera, G., Clouston, S., Reynolds, C. A., Thorvaldsson, V., Deary, I. J., et al. (2013). Coordinated analysis of age, sex, and education effects on change in MMSE scores. *Journals of Gerontology. Series B, Psychological Sciences and Social Sciences, 68*, 374–390.

Shipley, B. A., Weiss, A., Der, G., Taylor, M. D., & Deary, I. J. (2007). Neuroticism, extraversion, and mortality in the UK health and lifestyle survey: A 21-year prospective cohort study. *Psychosomatic Medicine, 69*, 923–931.

Simmons, J. P., Nelson, L. D., & Simonsohn, U. (2011). False-positive psychology: Undisclosed flexibility in data collection and analysis allows presenting anything as significant. *Psychological Science, 22*, 1359–1366.

Stephan, Y., Sutin, A. R., Canada, B., & Terracciano, A. (2017). Personality and frailty: Evidence from four samples. *Journal of Research in Personality, 66*, 46–53.

Sudlow, C., Gallacher, J., Allen, N., Beral, V., Burton, P., Danesh, J., et al. (2015). UK biobank: An open access resource for identifying the causes of a wide range of complex diseases of middle and old age. *PLoS Medicine, 12*(3), e1001779.

Turiano, N. A., Chapman, B. P., Gruenewald, T. L., & Mroczek, D. K. (2015). Personality and the leading behavioral contributors of mortality. *Health Psychology, 34*(1), 51.

Turiano, N. A., Graham, E. K., Weston, S. J., Booth, T., Harrison, F., James, B. D., Lewis, N. A., Makkar, S., Mueller, S., Wisniewski, K. M. et al. (2019). *Is healthy neuroticism associated with longevity? A coordinated integrative data analysis*. Manuscript under review.

Turiano, N. A., Mroczek, D. K., Moynihan, J., & Chapman, B. P. (2013). Big 5 personality traits and interleukin-6: Evidence for "healthy neuroticism" in a us population sample. *Brain, Behavior, and Immunity, 28*, 83–89.

Turiano, N. A., Whiteman, S. D., Hampson, S. E., Roberts, B. W., & Mroczek, D. K. (2012). Personality and substance use in midlife: Conscientiousness as a moderator and the effects of trait change. *Journal of Research in Personality, 46*, 295–305.

Vartanian, T. P. (2010). *Secondary data analysis*. Oxford: Oxford University Press.

Vollrath, M., & Torgersen, S. (2002). Who takes health risks? A probe into eight personality types. *Personality and Individual Differences, 32*, 1185–1197.

Weston, S. J., Graham, E. K., Turiano, N. A., Aschwanden, D., Booth, T., Harrison, F., James, B. D., Lewis, N. A., Makkar, S., Mueller, S. et al. (2019). *Is healthy neuroticism associated with chronic conditions? A coordinated integrative data analysis*. Manuscript under review.

Weston, S. J., Hill, P. L., Edmonds, G. W., Mroczek, D. K., & Hampson, S. E. (2018). No evidence of "healthy neuroticism" in the Hawaii personality and health cohort. *Annals of Behavioral Medicine, 53*, 426–441.

Weston, S. J., & Jackson, J. J. (2015). Identification of the healthy neurotic: Personality traits predict smoking after disease onset. *Journal of Research in Personality, 54*, 61–69.

Weston, S. J., & Jackson, J. J. (2016). How do people respond to health news? The role of personality traits. *Psychology & Health, 31*, 1–18.

Weston, S. J., Ritchie, S. J., Rohrer, J. M., & Przybylski, A. K. (2019). Recommendations for increasing the transparency of analysis of pre-existing datasets. *Advanced Methods and Practices in Psychological Science, 2*(3), 214–227.

Yarkoni, T., & Westfall, J. (2017). Choosing prediction over explanation in psychology: Lessons from machine learning. *Perspectives on Psychological Science, 12*, 1100–1122.

Chapter 7
Using Ambulatory Assessments to Understand Personality-Health Associations

Joshua J. Jackson and Emorie D. Beck

The relationship between personality and health is well established. Assessments of short and simple personality measures are prospectively associated with objective outcomes of health such as stroke, Alzheimer's, heart disease, cancer, diabetes – just to name a few (Terracciano et al. 2014; Weston, Hill, & Jackson 2015). Personality associations with health outcomes, in turn, help explain the relationship between personality and mortality (e.g., Jackson, Connolly, Garrison, Levine, & Connolly 2015). Further, personality-health processes extend beyond personality as a prospective risk factor whereby personality also influences perceptions of health and prognoses of disease after onset (Weston & Jackson 2016). These associations are found across different time periods for when personality is assessed – from childhood through old age – and across assessment methods (Friedman & Kern 2014). In short, personality arguably influences aspects of the health process from cradle to grave.

These well-established associations between personality and health are explained by a number of potential pathways. One of the most studied is health behaviors, such that those high in certain traits are more likely to engage in healthy behaviors, like eating their vegetables, exercising, going to the doctor, and less likely to engage in risky activities such as drug use and driving recklessly (Bogg & Roberts 2004). Other pathways linking personality with health outcomes are stress mediating processes and interpersonal relationships. For the former, certain traits are linked with experiencing more stress and a greater likelihood of encountering stressful situations (Williams, Smith, Gunn, & Uchino 2011). These stress related processes have well documented relationships with a number of biological mechanisms such as impacting the glucocorticoid system.

J. J. Jackson (✉) · E. D. Beck
Department of Psychological and Brain Sciences, Washington University in St. Louis, St. Louis, MO, USA
e-mail: j.jackson@wustl.edu

P. L. Hill, M. Allemand (eds.), *Personality and Healthy Aging in Adulthood*, International Perspectives on Aging 26,
https://doi.org/10.1007/978-3-030-32053-9_7

Despite these straightforward and sensible pathways linking personality and health, evidence for these pathways are not robust: in general, the overall magnitude of association that mediated pathways explain is relatively small, the mediating processes are not well defined, and the designs of studies typically do not account for reverse causation (i.e. health influences personality; Jackson, Weston, & Schultz 2017). As a result of the lack of strong consistent empirical findings, there are few people outside of personality highlighting the importance of personality traits for health, with no major public policy pushes or discussions of personality interventions to overcome health crises (c.f. Bleidorn et al. 2019; Lahey 2009). The relatively small magnitude of association between health and mediating mechanisms is understandable when considering there are often decades between assessments of personality and health, as well as the cumulative, multi-determined nature of health, not to mention the difficulty of intervening on personality (Jackson, Beck, & Mike in press).

Personality-health pathways are undoubtably complex. However, these complex pathways linking personality and health are studied relatively simplistically, as most studies tend to look at static markers of personality, health behaviors and health outcomes, often cross-sectionally. However, the processes driving personality-health associations unfold across time and in different situations, calling into question the utility of cross-sectional designs. To truly understand personality-health associations requires broadening when, where, and how personality is assessed. This is accomplished by measuring personality processes repeatedly – in the situations in which they naturalistically occur – and viewing personality through a different lens. We echo the early sentiments of the "father" of personality psychology Gordon W. Allport that, "Novel and somewhat daring methods will be required" (Allport 1937, p. 20) and propose that novel design and assessment methods can open up new possibilities for better understanding the relationship between personality and health processes.

The current chapter describes how Ambulatory Assessments (AAs, also known as Experience Sampling Method (ESM), Ecological Momentary Assessments (EMA), or intensive longitudinal designs (ILD)), which result in multivariate time series data, can move the study of personality and health forward. In these types of designs, a set of variables from a single individual is assessed multiple times within or across days or weeks. While data such as these have been collected since the 1940s or earlier (e.g. Cattell 1957), there was no coherent framework on how to assess personality and such data were not routinely collected until the introduction of electronic assistants that help "ping" participants throughout the day. Since these electronic assistants has emerged, personality psychologists have used AA to investigate between- versus within-person sources of variability in personality (e.g. Fleeson 2001; Fleeson & Gallagher 2009; Sherman, Rauthmann, Brown, Serfass, & Jones 2015) and carryover-effects of personality states (autocorrelative relationships; Beck & Jackson 2019a, 2019b). However, there has been little discussion about how these types of designs can advance more applied personality questions, such as personality and healthy aging. That said, adjacent fields have routinely used

AA methods to better understand health-relevant aging processes within psychopathology (e.g., Trull & Ebner-Priemer 2013) and emotion (Conner & Barrett 2012).

Ambulatory assessments provide a snapshot of the person as embedded within their lived environment, providing more insight into the processes that connect personality with health, and sidestepping many issues that may arise with retrospective reports and/or generalized trait assessments. Not only do ambulatory assessments allow personality assessments to become contextualized and dynamic, they also allow rich numbers of repeated measures to better assess within-person relationships. Moreover, with the additional time series assessments of health via wearables, such as heart rate and activity, there are a multitude of opportunities to examine novel relationships between personality and health. Below we outline two ways by which AA methods can advance the study of personality and healthy aging.

7.1 Additional (Dynamic) Metrics of Personality Traits

Existing personality-health associations rely almost exclusively on relatively static and simplistic assessments of personality that have important shortcomings with regard to understanding personality processes. Personality assessments are typically self-reported and constitute global evaluations of how a person behaves, thinks, and feels. While alternative rating sources, such as through close associates, provide novel information in the prediction of health (Jackson et al. 2015), broad assessments of personality tend to provide a rough snapshot of a person that do not wholly convey the complexity of a person. Further, a number of personality traits have been implicated in both health promotion and health suppression. For example, neuroticism is generally negatively associated with health outcomes (Weston et al. 2015). In contrast, there are some potential beneficial aspects of neuroticism, too (Friedman 2000; Weston & Jackson 2018; Weston & Jackson 2016, though see Weston, Hill, Edmonds, Mroczek, & Hampson 2018). To disentangle the beneficial and negative aspects of neuroticism, one needs to examine what a person is like *when* they are enacting health behaviors. For that, we need to look at more intensive assessments, often involving a systematic assessment of time to better understand the directionality of association (e.g. do health behaviors better predict neuroticism or vice versa?). Each of these additional types of assessments described below may be related to healthy aging over and above typical mean-level assessments of personality.

7.1.1 Validating Self-Reports

Personality assessments were derived from questionnaires aimed to assess relatively stable attributes. Items are not contextualized within a particular experience, but instead are general descriptors of how someone behaves collapsed across any systematic variability in time and situations. The relatively stable component of personality traits assessed by global reports is likely responsible for the relationship between personality traits and important life outcomes. The cumulative aspects of health processes, where personality-behavior associations add up over time, depends on consistency of personality (e.g., smoking once likely will not have an effect, but it may over a lifetime). However, it is unclear whether global assessments of personality correspond to the lived life of a person as few studies have examined personality manifestations in daily behavior. AA methods can assist in validating self-report and other-report methods of personality to make sure that those who state they are conscientiousness do in fact behave conscientiously (Fleeson & Gallagher 2009).

7.1.2 Variability in Personality

Within-person variability in personality has been discussed since Allport (1937), and figures prominently in many different theories of personality (e.g., CAPS; Mischel & Shoda 1995). However, few people have looked to see whether or not people are variable, if that variability is systematic, and if variability differs across people (see Beck & Jackson 2019a for a further discussion) – let alone if that variability is related to health processes. Variability in personality is especially important for personality-health associations given that many health behaviors may occur because of shifts in one's personality state. For example, someone may be generally conscientiousness but sometimes lack self-control due to fatigue or boredom. This, in turn, could lead to a greater likelihood of an unhealthy behavior (Tsukayama, Duckworth, & Kim 2012). Those people that are more variable in personality may be more susceptible to these unhealthy behaviors compared to those that have equal trait standings, but less variability given that those who are more variable will be more likely to experience personality states that depart from their average levels.

Further, variability in personality manifestations may be useful in such that it indexes (usually operationalized through a simple within-person standard deviation in personality states) an ability to respond to the demands of the environment. This ability to shift one's personality manifestation from situation to situation may be beneficial in keeping with what is expected, whereas rigid behavior regardless of the situation could be seen as maladaptive (Mancini & Bonanno 2009). For example, there has been some work showing that abstaining from alcohol may be detrimental for one's health. Here, abstaining may be seen as being inflexible and too rigid in behavior that leads to negative consequences. It is possible that this effect stems from the personality of those that abstain, rather than the act of abstaining (Walton & Roberts 2004), such that those who abstain may have fewer social contacts due to perceived rigidity. Alternatively, variability from situation to situation has been described as a risk factor for poor mental health and greater distress (Hardy & Segerstrom 2017; Wichers, Wigman, & Myin-Germeys 2015), presumably resulting from the inability to have a consistent self. Currently, however, the extent that variability in personality states influences health is unknown.

Cattell (1957) broke down within-person variability into *oscillations* (momentary changes) and *fluctuations* (changes over longer time periods, like days or weeks). As applied to personality-health associations, it is likely that stressful times in life prompt people whose personality profiles may put them at risk to engage in less healthy behavior, such as those high on neuroticism, may be more likely to be influenced by outside stressors. The incident can be relatively acute, such as a hectic day at the office that results in poor food choices and a neglect of daily activity, stemming from exhaustion and the need to relax. The result is a temporary oscillation of personality-health behavior associations, one that may or may not have long-term consequences. However, this association could also be prolonged, leading to long term fluctuations in personality states. It is possible that neurotic individuals, to extend the example, may have more of these fluctuation episodes across the lifespan where they demonstrate high levels of neuroticism states. It is the persistence and the accumulation of these episodes that contributes to poorer health down the road – as opposed to what tends to occur day-to-day outside these fluctuations. In such cases, the experience of these fluctuations may be reinforced, for example, by the reduction of negative affect that follows eating unhealthy food or neglecting daily activity, making those behaviors more likely in the future when neuroticism fluctuations (inevitably) occur. That is, a simple oscillation of being stressed out probably isn't going to have long term ramifications for health. Longer term shifts, fluctuations, may be what is driving the association between personality and health. Examining only general, trait level personality-health associations will thus miss these important personality – health transactions.

Another related concept is *instability*, which is how extreme deviations are when they occur. This captures not just how much a person varies but also how extreme or gradual they are when they do occur. Whereas variability indices like standard deviation captures deviations around the mean, instability reflects the temporal dependency of fluctuations in states over time. Instability is often measured through mean square successive differences (MSSDs) which tallies the squared difference from one time to the next. When considering neuroticism, whose positive pole is emotional stability, a measure of instability is a valuable additional metric in addition to within-person variability alone. For example, two individuals could show identical within-person variability (average scaled difference between an observation and the average observation; SD) but very different MSSDs. If, across 20 observations, one individual gave the same rating (e.g. "1") to the first half of observations and a different rating (e.g. "2") to the second half of observations, while the other individual alternated between the two ratings for the whole period, they would have the same within-person variability (SD = .51) but different MSSD's ($MSSD_1$ = .05; $MSSD_2$ = 1).

Changing states in personality from one situation or time to another, as discussed above, could be beneficial or harmful. There is some evidence that instability influences health, coming from the affective instability literature. Given that negative and positive affect relate closely to neuroticism and extraversion, respectively, it is likely that these findings would extend to personality fluctuations (Ong & Ram 2017). More research needs to be done to understand if and how personality instability is related to health. For example, is good or poor health that leads to more or less instability or does personality instability influence health? Those that are more variable from time-point to time-point likely are, almost by definition, more susceptible to situational influences. Thus, those people who are high in instability might be prime targets for behavioral interventions as their manifestations appear more labile.

7.1.3 Inertia

Like in physics, inertia refers to the tendency to stay consistent within a particular state. Often, inertia is defined as the autoregressive correlation from one time to the next. Similar to instability, inertia involves temporal dependency, though it does not index the magnitude of change from time-point to time-point. Inertia is relatively high within personality panel studies that examine personality from year-to-year, but fewer studies have investigated this within daily or more fine-grain assessments of personality (c.f. Beck & Jackson 2019a; Epskamp, Waldorp, Mõttus, & Borsboom 2018 for exceptions). In these studies, it is relatively unsurprising that state manifestations of personality show modestly high auto-correlations. However, there are large individual differences in the tendency to show high auto-correlations. No personality studies have investigated whether individual differences in inertia are related to different health outcomes. Outside of the trait literature, very few studies have looked at inertia as it relates to health relevant processes (e.g., Kuppens, Allen, & Sheeber 2010; van de Leemput et al. 2014). One study looked at the inertia of positive affect found that those low in inertia had lower levels of depressive symptoms, indicating that having greater stability in your personality states was positive (Hohn, Menne-Lothmann, Peeters, et al. 2013).

Inertia is typically discussed as the autocorrelation of the same variable across time e.g., state levels of extraversion. However, it can be extended to so-called cross-lagged associations – looking at how one's extraversion predicts future levels of conscientiousness, for example. These cross-lagged associations have not been examined yet either in terms of what is normative or how they relate to health processes. However, inertia across constructs can be viewed in terms of *if…then* associations that are the basis for many process models of personality (Mischel & Shoda 1995; Wright & Mischel 1987). For example, if someone was feeling depressed, a facet of neuroticism, then that may make it less likely for them to behave conscientiously, which may manifest in skipping working out to eat junk food. In contrast, someone who has lower levels of inertia across these two constructs may be less likely to exhibit these poor health behaviors because their elevation in neuroticism does not influence conscientious behaviors. These people may be able to shrug off or compartmentalize whatever left the person feeling depressed more so than the person with high inertia. The extent that individual differences exist for cross-lagged inertia in personality is unknown.

7.1.4 Cyclic Trends

Periodicity, or time dependent patterns in changes in a state over time, of psychological states have been largely unaddressed in personality psychology. Despite that, these cyclic trends, which could operate on the level of minutes and hours to days and months, can be easily examined in AA studies. Moreover, cycles are known to be important for physical functioning, impacting, for example, sleep and eating processes. Separating cyclic trends characteristic from autocorrelative associations is important in understanding both those effects. For example, there are clear diurnal cycles in energy levels, which is often a measure of the activity level facet of extraversion, across the day. If cyclic trends are not accounted for, this may mask patterns of variability *within* a day by not accounting for variability across days. In a demonstration of how cyclical trends are critical. in understanding the processes that underlie personality, Revelle and colleagues (1980) examined arousal levels as a function of extraversion and diurnal cycles, finding that higher levels of arousals among those low in extraversion occurred only in the morning. Indeed, the cyclic trend itself may be an individual difference characteristic (e.g. the periodicity in arousal may be predictive of both personality traits and health outcomes). These issues are especially important in aging context as people's energy levels and diurnal cycles change with age and health status (Froy 2013). Combined with the changes in extraversion found in older adulthood (Wortman, Lucas, & Donnellan 2012), there may be large differences in personality-health associations in older adults compared to younger.

7.1.5 Reactivity

Outside events are likely to influence one's personality manifestations, often thought of as person-by-situation interactions. These events, however, may be more or less impactful depending on the person. This idea is sometimes thought of as resiliency when the time frame is longer than moment to moment. But from moment to moment, people may also be more or less susceptible to outside influences, which could directly have implications for health. Reactivity in positive affect in the face of daily pain was associated with vulnerability in fibromyalgia patients (Finan, Zautra, & Davis 2009). Similarly, people whose positive affect were impacted more by daily stressors have poorer sleep and elevated inflammation (Ong, Exner-Cortens, Riffin, et al. 2013; Sin, Graham-Engeland, Ong, & Almeida 2015). While personality x situation interactions are often discussed, within the personality literature, few empirical studies can reliably show an interaction between the two (e.g. Sherman et al. 2015). It is possible that finer-grained AA studies could shed light on how people react to outside influences.

In sum, few studies have systematically utilized additional AA derived metrics of personality to better understand associations with health. However, merely introducing these ideas into the personality-health literature will not be satisfactory. More systematic research is needed to best design intensive longitudinal assessments, including what and with what frequency. Many of these metrics would change if the time period between assessments changed, for example. For some questions, daily repeated assessments may be more useful than hourly, whereas the reverse may be the case for others. This is further complicated by the potential to assess meaningful health metrics continuously through various wearables (e.g., heart rate). Relatedly, there have been few if any systematic psychometric evaluations of personality or health measures that are used within AA studies. It is not immediately clear that translating a trait questionnaire to a state questionnaire will result in a well validated scale. Moreover, there are additional considerations concerning validating AA indicators (Wright & Zimmerman 2019), especially when considering that many AA studies address within-person questions compared to standard between person that measures were traditionally validated against.

7.2 Idiographic Assessments

The time series data collected through AA allows new variables to be created, like those described above, but also allows new opportunities for conceptualizing personality. Traditional personality assessments are nomothetic in nature such that they assume that all people can be characterized by similar constructs – that is, (1) my extraversion is the same as your extraversion and (2) we can then rank each other on levels of extraversion. An alternative approach to this is idiographic assessment where personality is interpreted as the relative standing of variables within a person, such that not all variables will be important for all people. These assessments are more akin to how you would describe someone after knowing them really well, through discussion of components of their personality that are especially salient to them, not just through listing their relative standings on each of the Big Five traits. The Big Five has been referred to as the "psychology of the stranger" (McAdams 1994), referring to the rough approximation of personality for someone who barely knows a person. Once two people move past stranger status, a series of predefined constructs such as the Big Five may not be necessary as they are overly coarse. Instead, you can focus on the characteristics that are salient to an individual and how these characteristics are organized specific to an individual. It is likely that this increased precision could open up new insights into personality-health associations.

Idiographic assessment has been discussed since the days of Allport, Cattell, and others (Beck & Jackson 2019a). Allport, for example, talked about cardinal traits and individual traits, both of which go beyond typical nomothetic assessments of personality (Allport 1937). Cardinal traits are traits that are especially relevant for a person in their daily life. The workaholic business executive may list industriousness as their cardinal trait whereas the comedian may feel that humor is the most salient trait that defines her actions. Individual traits go one step further in that both the structure and the content of your personality may only exist within you and not apply to any other person (Allport 1937). To Allport, personality exists at two levels – at the level of the individual (idiographic) and the level of the population (nomothetic).

Cattell (1957) proposed how to think about these two levels through his databox where persons could be conceptualized into three dimensions that indexed people (P_1 to P_N), variables (X_1 to X_p), and occasions or time (T_1 to T_t). Depending how you slice up the data box, you could ask nomothetic questions that applied to all people or idiographic questions that applied to a single person. Nomothetic questions examine the person (P) and variable (X) dimensions and collapse across the occasion (T) dimension either implicitly or explicitly. With this slice, one can examine the nomothetic structure of individual differences within a population of people, which he termed *R-technique*. Instead of collapsing across time, one could slice the data box into variable (X) and occasion (T) dimensions and fix the person dimension. In doing so one can capture the unique structure of individual differences for a single person – that is, the within-person idiographic factor structure of personality. Cattell described this procedure as *P-technique*.

Despite the introduction of analytic tools to examine both between and within-person questions over 60 years ago, few studies have addressed questions of idiographic personality. Some of the progress was limited by the difficulty in collecting the data. Similarly, even with good AA data, computing power limited the types of questions one could address. Finally, the Person-Situation Debate that waged on throughout the latter half of the twentieth century pitted between-person nomothetic perspectives and within-person idiographic perspectives against each other. The result is that there were few people interested in crossing party lines to investigate how thinking about personality at these different levels could be adequately addressed. Given this disconnect, it is necessary to connect the two sciences of personality (idiographic and nomothetic) to determine the best level of analysis for a specific question. In terms of aging and health, there may be general processes that drive between person associations such that feeling stressed is associated with worse health outcomes. However, this may not be true for everyone. Stress for some may not be associated with negative health processes but with positive health promoting processes. Only looking for processes at the between-person level may thus overlook important within-person (idiographic) processes.

7.2.1 Distinguishing Nomothetic and Idiographic Processes

The novel personality metrics discussed above (inertia, instability) that address the dynamics of personality over time are based on between-person models, like factor analysis and most longitudinal models. These models assume that whatever characteristics one is looking at, such as the inertia of extraversion, exists similarly for everyone. However, translating the findings of between-person models to individual, idiographic models often provides conflicting results such that the same associations do not hold at the between- versus the within-person level (Borsboom, Mellenbergh, & Van Heerden, 2003; Molenaar 2004; Fisher, Medaglia, & Jeronimus, 2018). Mathematically, if within-and between-person processes were equivalent, they would be *ergodic* (Molenaar 2004). In psychology, this would mean that there would be equivalence between P-technique (X x T, fixing P) and R-technique (X x P, fixing T). Comparisons of between and within person mood processes, sometimes find equivalent structures for most participants – but not all (Zevon & Tellegen 1982). In a recent demonstration, Fisher et al. (2018) investigated the equivalence of within-and between-person means and standard deviations, finding that the distributions were quite different, which indicates that the time series are non-ergodic and individual-level estimates cannot be made from between-person estimates.

If mismatch between idiographic and nomothetic models is strong, as preliminary data within personality suggest (e.g. Beck & Jackson 2019a; Borkenau & Ostendorf 1998; Molenaar 2004), then psychologists must turn their attention back to testing idiographic questions posed by Allport and others almost a century ago. This is because, what is considered "conscientiousness" may not exist in (1) all people or (2) in the same way. Which raises issues for using such nomothetic traits as causal explanations. Discussing how conscientiousness predicts mortality could thus be mingling different underlying processes that give rise to similar behaviors. In other words, my conscientiousness may not be the same as your conscientiousness at such a level that it doesn't make sense to talk about someone having a level of conscientiousness. Recently, for example, Beck and Jackson (2019a) examined the congruence between idiographic and between-person personality structure using well-known (if not oft applied) techniques, like P-factor analysis, as well as newer models that account for timing (see below). Across each method, they found little congruence between the between-person reference structure and idiographic structures, with some people showing strong congruence (a similar factor structure) but many showing weak congruence at best. That is, people differ in the number of personality factors, with the content of those factors not necessarily replicating either. In short, the personality structure of an individual does not necessarily map on to the current taxonomies of personality structure (e.g., HEXACO; Big Five), with some people not having factors that resemble any of the standard traits. People are unique in how their personality is organized, with some components being important for some people but not others.

Although further work is needed to better understand the correspondence between within- and between-person models of personality structure, and how variability can be leveraged to improve our understanding, this study represents the most comprehensive study to date examining methods for assessing congruence. The lack of evidence for such congruence is striking and further highlights the need for devoting more attention to idiographic assessment and modeling. If personality traits are causally influencing distal outcomes decades down the road, then it is necessary to identify the components that are the influencers. The findings from Beck and Jackson (2019a) indicate it is incorrect to label such terms like conscientiousness as distal influences (e.g., Hill & Jackson 2016; Jackson & Hill 2018) given that conscientiousness may not "exist" for some individuals.

7.2.2 *Idiographic Conceptualization of Personality Traits*

Theoretically, critical assumptions of the factor model underlying the derivation of nomothetic (common) traits in personality make it unlikely that nomothetic structures can be used to understand individual-level dynamics. Models of the Big Five were originally extracted using exploratory factor analysis, often with Procrustes rotations that favor its extraction. Less restrictive assessments of the factor structure across the lifespan suggest that the Big Five factor structure do not hold up (Beck, Condon, & Jackson 2019). The factor structure is replicable in early adulthood, the same age that initial validation studies tended to use, but in older ages there was less consistency in the factor structure, suggesting de-differentiation among people. In other words, in older adulthood, personality got more complex with greater links between variables (i.e. cross loadings). What this means is that findings based on factor scores of older adults may be either driven by a small subset of people where the trait exists for them and/or that our current estimates for the effect sizes between personality and life outcomes are impacted because our assessments of personality are imprecise. Both possibilities are chilling if personality researchers want to identify the mechanisms between personality and healthy aging. Regardless, these findings suggest that standard factor models of personality may not apply to everyone and that research into healthy aging should utilize idiographic assessment more frequently.

Three important assumptions of the standard *reflective model* (or common effect model; see Fig. 7.1) used in nomothetic personality trait studies are that (1) the indicators of these traits are assumed to be interchangeable i.e. that they all measure the same construct, (2) their errors are independent after accounting for the latent variable, meaning that only the latent variable explains variability in these indicators, not other variables, and (3) the latent trait causes variability in the indicators and is not just a data reduction technique. The use of the reflective model in personality has been criticized for violating each of those three assumptions (e.g. Cramer et al. 2012). Although the independence and interchangeability of indicators assumptions can be somewhat bypassed by fitting confirmatory factor models that

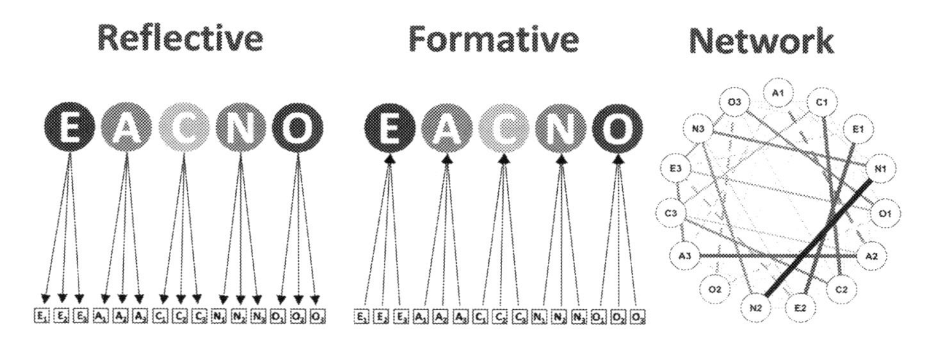

Fig. 7.1 Three potential models for conceptualizing personality structure

relax these assumptions (e.g. Hopwood & Donnellan 2010), the question of the causal nature of latent variables remains. However, there are other models that are mathematically equivalent to the common reflective model that relax all of these assumptions (Kruis & Maris 2016). Despite their mathematical equivalence, each of the models proposes a different theoretical explanation for the observed associations. Thus, the choice of the model has critical implications for a theory about how personality influences outcomes such as health.

Two other models that are mathematically equivalent to the reflective model – the *formative model* (or common cause model) and an *idiographic network model* (Fig. 7.1). In a formative model, a latent variable is regressed on its indicators. A classic example of this is socioeconomic status, where indicators like education, household income, and location are indicators of the construct SES. However, there is no assumption that education, income, or location *cause* SES. In personality, in other words, observable manifestations of Extraversion, like being talkative, are not caused by some Extraversion trait that a person has. Rather, their position on the latent construct Extraversion is a function of their tendency to be talkative, as well as a number of other indicators. However, although the formative model relaxes assumptions of the reflective model that often do not hold (perhaps especially the independence of errors), it fails to provide a satisfying explanation for why latent traits emerge at all.

The idiographic network model, though, not only relaxes reflective model assumptions but also suggests explanations for why latent trait dimensions emerge. Based in dynamic systems theory, an idiographic network model asserts that latent traits emerge as emergent properties of interactions among a set of indicators. These emergent properties may appear to be "trait like" but they are only a property of the complex interrelations among lower order components. As a result, these models make none of the assumptions of the common, reflective model. In doing so, these models offer up new perspectives on how to assess and capture personality processes (Beck & Jackson 2017).

Instead of focusing on the causal properties of broad traits, lower order components are assumed to be causally responsible for personality-health associations. By looking at these lower order components, compared to standard trait models, idiographic models introduce more unique personality → health pathways, allowing for more complex models that capture personality-health associations in the complexity that many theories assert. For example, Idiographic network models allow one to systematically estimate the associations between indicators and target behaviors for each person. One can then ask whether (1) properties of the structure of someone's personality (not their trait standing) is associated with health? Or (2) whether a specific pathway is found (e.g., feeling stressed results in more or less than desired exercise) – and does that association hold for everyone? Currently questions of how idiographic personality structure and network processes relate to health processes is completely unknown.

7.2.3 Modern Idiographic Network Models

Idiographic network models of personality differ from standard trait assessments in that they require AA data. They shift the focus from static trait assessments to more dynamic state-level assessments of personality for which AA methods allow. The simplest type of these models is a zero-order correlation among the variables (X) dimension across the time (T) dimension for each person individually, directly echoing Cattell's slicing of the databox. Called "association networks," these are the correlation matrices that P-technique factor analysis attempts to reduce. By providing an intuitive visualization of the relationships among indicators, network techniques provide an easy tool for visualizing structural differences in personality traits.

These methods can be expanded to looking at unique relations among indicators and can utilize the time sensitive nature of AA data to looked at lagged associations. However, when examining such within and across (lagged) personality-health relationships in the search for casual relationships strong overlapping variance among indicators can threaten this endeavor, making it important to examine the unique relationships among the indicators using partial correlations or controlling via regression. As with standard regression, the possibility of over-controlling or overfitting goes up as the model increases in complexity (assuming the number of observations remains constant). Recently, there have been a number of proposed models for examining such complex models without overfitting. New techniques for the basic lagged model (e.g. Bringmann et al. 2016; Epskamp et al. 2018; Gates & Molenaar 2012; Wild et al. 2010), have been proposed and implemented (in a limited manner) to account for (1) bidirectional relationships between lagged predictors, (2) additive relationships among lagged predictors, and (3) the structure of these relationships (Beck & Jackson 2019a). Together, these novel analytic models allow researchers to address the complex analytics that come with AA data, especially as it applies to idiographic assessment and to tackle questions that have long been of interest.

Modeling personality as cross-lagged idiographic models have several advantages. First, most types of models account both for within- (contemporaneous) and across- (lagged) time relationships. This is important, as personality-health associations likely differ in the time scale of the associations. For example, my anxiety right now may drive my current eating habits whereas my planfulness in the moment will impact future ability to exercise. Second, by including larger set of predictors and using pruning techniques to prevent multicollinearity (e.g. graphical LASSO; Friedman, Hastie, & Tibshirani 2008), they capture the unique relationships among diverse phenomena that influence manifestations of personality. That is, these models can identify the relevantly salient paths that are most likely causal by accounting for overlapping variance, much like control variables do so in standard regression models. Third, and most simply, they offer a method for testing complex sets of relationships that are a hallmark of many key models of personality (Allport 1937; Cervone 2005; Mischel & Shoda 1995). Such relationships are complex not only in that they can include a large number of predictors but also in what those predictors are.

7.2.4 Implications for Healthy Aging

These newer idiographic models can provide a number of insights about personality structure, personality processes, and personality development (Beck & Jackson 2017) that may be especially relevant to health processes. For example, while conscientiousness is known to be associated with better health, it is possible that two individuals who have identical levels of Conscientiousness on a nomothetic personality scale obtain better health through different means. Traditionally, from a nomothetic view, this problem has been approached by examining lower-order personality characteristics, such as facets (Jackson et al. 2009; Soto et al. 2011), aspects (DeYoung, Quilty, & Peterson 2007), or items (Mõttus, Kandler, Bleidorn, Riemann, & McCrae 2017; Wood, Nye, & Saucier 2010) in an attempt to obtain more precision about the pathway. But this still misses unique associations *between* lower-order personality characteristics, such as whether an individual's sociability is related their impulsivity, which impacts health processes. Allowing for dependencies in idiographic models allows for novel pathways such as these to be tested that are not possible using traditional assessments.

Similarly, idiographic data are inherently temporal. As a result, idiographic models are able to better examine the effects of context and time and to examine so called *if…then* contingencies (Wright & Mischel 1987) that likely play an important role in personality-health associations (Weston & Jackson 2016). For example, compared to traditional models of personality, idiographic models incorporate multiple different situations due to repeated responses. As such, situations can be directly incorporated into the same model to see how personality influences health processes when surrounded by people versus not, for example. It is possible that a someone's personality could have a completely different structure and thus association with health relevant variables across different contexts. If pathways related to health (risky driving, food choices) predominantly exist in one of those domains, it is important to assess the personality structure in that domain, not average across domains or as assessed in another. In contrast, more traditional approaches to integrating person and situations must statistically interact separate personality and situation terms (e.g. Sherman et al. 2015), which may miss idiosyncratic patterns of person-situation transactions that idiographic network models can address.

7.3 Conclusion

Personality is a powerful predictor of health, but there are few well-established pathways between personality and health outcomes. The current chapter described two broad techniques to better characterize personality in the hopes of uncovering these pathways. Despite the promise, dynamic metrics and idiographic methods have not been applied to personality and health data. For the former, there is some evidence that mean levels are better predictors of well-being than the more dynamic metrics described above (Dejonckheere et al. 2019); however, it remains unclear whether this is also the case for personality health-related outcomes. For the latter, a common misconception is that it is hard to describe general tendencies or make broad conclusions if every person in your sample is unique. Both of these perceived limitations are unfounded. No study has examined dynamic metrics of personality to related them to health. Idiographic models can be described as distributions of parameter estimates or as classes of similar people. Together we feel that moving personality assessment beyond simple mean averages can inject new blood into the study of personality and health.

References

Allport, G. W. (1937). *Personality: A psychological interpretation.*

Beck, E.D., Condon, D., & Jackson, J.J. (2019). *Interindividual age differences in personality structure*. Manuscript in preparation.

Beck, E. D., & Jackson, J. J. (2017). The search for a bridge: Idiographic personality networks. *European Journal of Personality, 31*, 530–532.

Beck, E. D., & Jackson, J. J. (2019a). Consistency and change in idiographic personality: A longitudinal ESM network study. *Journal of Personality and Social Psychology*. https://doi.org/10.1037/pspp0000249

Beck, E. D., & Jackson, J. J. (2019b). Within person variability. In J. F. Rauthman (Ed.), *Handbook of personality dynamics and processes*. New York, NY: Academic.

Bleidorn, W., Hill, P., Back, M., Denissen, J. J. A., Hennecke, M., Hopwood, C. J., … Roberts, B. (2019). The policy relevance of personality traits. *American Psychologist*. https://doi.org/10.31234/osf.io/a9rbn

Bogg, T., & Roberts, B. W. (2004). Conscientiousness and health-related behaviors: A meta-analysis of the leading behavioral contributors to mortality. *Psychological Bulletin, 130*(6), 887.

Borkenau, P., & Ostendorf, F. (1998). The Big Five as states: How useful is the five-factor model to describe intraindividual variations over time? *Journal of Research in Personality, 32*(2), 202–221.

Borsboom, D., Mellenbergh, G. J., & Van Heerden, J. (2003). The theoretical status of latent variables. *Psychological Review, 110*, 203.

Bringmann, L. F., Pe, M. L., Vissers, N., Ceulemans, E., Borsboom, D., Vanpaemel, W., … Kuppens, P. (2016). Assessing temporal emotion dynamics using networks. *Assessment, 23*(4), 425–435.

Cattell, R. B. (1957). *Personality and motivation structure and measurement.*

Cervone, D. (2005). Personality architecture: Within-person structures and processes. *Annual Review of Psychology, 56*, 423–452.

Conner, T. S., & Barrett, L. F. (2012). Trends in ambulatory self-report: The role of momentary experience in psychosomatic medicine. *Psychosomatic Medicine, 74*(4), 327.

Cramer, A. O., Van der Sluis, S., Noordhof, A., Wichers, M., Geschwind, N., Aggen, S. H., ... Borsboom, D. (2012). Dimensions of normal personality as networks in search of equilibrium: You can't like parties if you don't like people. *European Journal of Personality, 26*(4), 414–431.

Dejonckheere, E., Mestdagh, M., Houben, M., Rutten, I., Sels, L., Kuppens, P., & Tuerlinckx, F. (2019). Complex affect dynamics add limited information to the prediction of psychological well-being. *Nature Human Behaviour, 3*(5), 478.

DeYoung, C. G., Quilty, L. C., & Peterson, J. B. (2007). Between facets and domains: 10 aspects of the big five. *Journal of Personality and Social Psychology, 93*(5), 880.

Epskamp, S., Waldorp, L. J., Mõttus, R., & Borsboom, D. (2018). The Gaussian graphical model in cross-sectional and time-series data. *Multivariate Behavioral Research, 53*(4), 453–480.

Finan, P. A., Zautra, A. J., & Davis, M. C. (2009). Daily affect relations in fibromyalgia patients reveal positive affective disturbance. *Psychosomatic Medicine, 474–482*(33), 71.

Fisher, A. J., Medaglia, J. D., & Jeronimus, B. F. (2018). Lack of group-to-individual generalizability is a threat to human subjects research. *Proceedings of the National Academy of Sciences, 115*(27), E6106–E6115.

Fleeson, W. (2001). Toward a structure-and process-integrated view of personality: Traits as density distributions of states. *Journal of Personality and Social Psychology, 80*(6), 1011.

Fleeson, W., & Gallagher, P. (2009). The implications of Big Five standing for the distribution of trait manifestation in behavior: Fifteen experience-sampling studies and a meta-analysis. *Journal of Personality and Social Psychology, 97*(6), 1097.

Friedman, H. S. (2000). Long-term relations of personality and health: Dynamisms, mechanisms, tropisms. *Journal of Personality, 68*(6), 1089–1107.

Friedman, H. S., & Kern, M. L. (2014). Personality, well-being, and health. *Annual Review of Psychology, 65*, 719–742.

Friedman, J., Hastie, T., & Tibshirani, R. (2008). Sparse inverse covariance estimation with the graphical lasso. *Biostatistics, 9*(3), 432–441.

Froy, O. (2013). Circadian aspects of energy metabolism and aging. *Ageing Research Reviews, 12*(4), 931–940.

Gates, K. M., & Molenaar, P. C. (2012). Group search algorithm recovers effective connectivity maps for individuals in homogeneous and heterogeneous samples. *NeuroImage, 63*(1), 310–319.

Hardy, J., & Segerstrom, S. C. (2017). Intra-individual variability and psychological flexibility: Affect and health in a National US sample. *Journal of Research in Personality, 69*, 13–21.

Hill, P. L., & Jackson, J. J. (2016). The invest-and-accrue model of conscientiousness. *Review of General Psychology, 20*, 141–154. https://doi.org/10.1037/t07564-000

Hohn, P., Menne-Lothmann, C., Peeters, F., et al. (2013). Moment-to-moment transfer of positive emotions in daily life predicts future course of depression in both general population and patient samples. *PLoS ONE, 8*, e75655.

Hopwood, C. J., & Donnellan, M. B. (2010). How should the internal structure of personality inventories be evaluated? *Personality and Social Psychology Review, 14*(3), 332–346.

Jackson, J. J., Beck, E. D., & Mike, A. (in press). Personality interventions. In John & Robins (Eds.), *Handbook of personality: Theory and research* (4th ed.). Washington, DC: HHS Public Access.

Jackson, J. J., Bogg, T., Walton, K., Wood, D., Harms, P. D., Lodi-Smith, J. L., & Roberts, B. W. (2009). Not all conscientiousness scales change alike: A multi-method, multi-sample study of age differences in the facets of conscientiousness. *Journal of Personality and Social Psychology, 96*, 446–459.

Jackson, J. J., Connolly, J. J., Garrison, M., Levine, M., & Connolly, S. L. (2015). Your friends know how long you will live: A 75 year study of peer-rated personality traits. *Psychological Science, 26*, 335–340. https://doi.org/10.1177/0956797614561800

Jackson, J. J., & Hill, P. L. (2018). Lifespan development of conscientiousness. In McAdams, Shiner, & Tackett (Eds.), *Handbook of personality across the Lifespan*. Berlin, Germany: Blackwell Publishing Ltd.

Jackson, J. J., Weston, S. J., & Schultz, L. (2017). Health and personality development. In J. Specht (Ed.), *Handbook of personality development*. New York, NY: Psychology Press.

Kruis, J., & Maris, G. (2016). Three representations of the Ising model. *Scientific Reports, 6*, 34175.

Kuppens, P., Allen, N. B., & Sheeber, L. B. (2010). Emotional inertia and psychological maladjustment. *Psychological Science, 21*(7), 984–991.

Lahey, B. B. (2009). Public health significance of neuroticism. *American Psychologist, 64*(4), 241.

Mancini, A. D., & Bonanno, G. A. (2009). Predictors and parameters of resilience to loss: Toward an individual differences model. *Journal of Personality, 77*(6), 1805–1832.

McAdams, D. P. (1994). A psychology of the stranger. *Psychological Inquiry, 5*(2), 145–148.

Mischel, W., & Shoda, Y. (1995). A cognitive-affective system theory of personality: Reconceptualizing situations, dispositions, dynamics, and invariance in personality structure. *Psychological Review, 102*(2), 246.

Molenaar, P. C. (2004). A manifesto on psychology as idiographic science: Bringing the person back into scientific psychology, this time forever. *Measurement, 2*(4), 201–218.

Mõttus, R., Kandler, C., Bleidorn, W., Riemann, R., & McCrae, R. R. (2017). Personality traits below facets: The consensual validity, longitudinal stability, heritability, and utility of personality nuances. *Journal of Personality and Social Psychology, 112*(3), 474.

Ong, A. D., Exner-Cortens, D., Riffin, C., et al. (2013). Linking stable and dynamic features of positive affect to sleep. *Annals of Behavioral Medicine, 52–61*(34), 46.

Ong, A. D., & Ram, N. (2017). Fragile and enduring positive affect: Implications for adaptive aging. *Gerontology, 63*(3), 263–269.

Sherman, R. A., Rauthmann, J. F., Brown, N. A., Serfass, D. G., & Jones, A. B. (2015). The independent effects of personality and situations on real-time expressions of behavior and emotion. *Journal of Personality and Social Psychology, 109*(5), 872.

Sin, N. L., Graham-Engeland, J. E., Ong, A. D., & Almeida, D. M. (2015). Positive and negative affective responses to daily stressors are associated with inflammation. *Health Psychology, 12*, 1154–1165.

Soto, C. J., John, O. P., Gosling, S. D., & Potter, J. (2011). Age differences in personality traits from 10 to 65: Big five domains and facets in a large cross-sectional sample. *Journal of Personality and Social Psychology, 100*(2), 330.

Terracciano, A., Sutin, A. R., An, Y., O'Brien, R. J., Ferrucci, L., Zonderman, A. B., & Resnick, S. M. (2014). Personality and risk of Alzheimer's disease: New data and meta-analysis. *Alzheimer's & Dementia, 10*(2), 179–186.

Trull, T. J., & Ebner-Priemer, U. (2013). Ambulatory assessment. *Annual Review of Clinical Psychology, 9*, 151–176.

Tsukayama, E., Duckworth, A. L., & Kim, B. (2012). Resisting everything except temptation: Evidence and an explanation for domain-specific impulsivity. *European Journal of Personality, 26*(3), 318–334.

van de Leemput, I. A., Wichers, M., Cramer, A. O., Borsboom, D., Tuerlinckx, F., Kuppens, P., … Derom, C. (2014). Critical slowing down as early warning for the onset and termination of depression. *Proceedings of the National Academy of Sciences, 111*, 87–92.

Walton, K. E., & Roberts, B. W. (2004). On the relationship between substance use and personality traits: Abstainers are not maladjusted. *Journal of Research in Personality, 38*(6), 515–535.

Weston, S., Hill, P. L., & Jackson, J. J. (2015). Personality traits predict the onset of major disease. *Social Personality Psychological Science, 6*, 309–317. https://doi.org/10.1177/1948550614553248

Weston, S. J., Hill, P. L., Edmonds, G. W., Mroczek, D. K., & Hampson, S. E. (2018). No evidence of "healthy neuroticism" in the Hawaii personality and health cohort. *Annals of Behavioral Medicine, 53*(5), 426–441.

Weston, S. J., & Jackson, J. J. (2016). How do people respond to health news? The role of personality traits. *Psychology & Health, 31*(6), 637–654. https://doi.org/10.1080/08870446.2015.1119274

Weston, S. J., & Jackson, J. J. (2018). The role of vigilance in the relationship between neuroticism and health. *Journal of Research in Personality, 73*, 27–34.

Wichers, M., Wigman, J., & Myin-Germeys, I. (2015). Micro-level affect dynamics in psychopathology viewed from complex dynamic systems theory. *Emotion Review, 7*, 362–367.

Wild, B., Eichler, M., Friederich, H. C., Hartmann, M., Zipfel, S., & Herzog, W. (2010). A graphical vector autoregressive modelling approach to the analysis of electronic diary data. *BMC Medical Research Methodology, 10*(1), 28.

Williams, P. G., Smith, T. W., Gunn, H. E., & Uchino, B. N. (2011). Personality and stress: Individual differences in exposure, reactivity, recovery, and restoration. In R. J. Contrada & A. Baum (Eds.), *The handbook of stress science: Biology, psychology, and health* (pp. 231–245). New York, NY: Springer.

Wood, D., Nye, C. D., & Saucier, G. (2010). Identification and measurement of a more comprehensive set of person-descriptive trait markers from the English lexicon. *Journal of Research in Personality, 44*(2), 258–272.

Wortman, J., Lucas, R. E., & Donnellan, M. B. (2012). Stability and change in the Big Five personality domains: Evidence from a longitudinal study of Australians. *Psychology and Aging, 27*(4), 867–874. https://doi.org/10.1037/a0029322

Wright & Zimmerman. 2019. *Applied ambulatory assessment: Integrating idiographic and nomothetic principles of measurement*. Psychological assessment.

Wright, J. C., & Mischel, W. (1987). A conditional approach to dispositional constructs: The local predictability of social behavior. *Journal of Personality and Social Psychology, 53*(6), 1159.

Zevon, M. A., & Tellegen, A. (1982). The structure of mood change: An idiographic/nomothetic analysis. *Journal of Personality and Social Psychology, 43*(1), 111–122. https://doi.org/10.1037/0022-3514.43.1.111

Chapter 8
Sounds of Healthy Aging: Assessing Everyday Social and Cognitive Activity from Ecologically Sampled Ambient Audio Data

Burcu Demiray, Minxia Luo, Alma Tejeda-Padron, and Matthias R. Mehl

As discussed in previous chapters, an important path connecting personality with healthy aging is via behavior. Individuals who have different personalities are likely to engage in different activities or engage in activities differently in everyday life. The different activity patterns of aging individuals may, over time, result in different health outcomes.

The World Health Organization (2015) recently proposed a healthy aging model. This model conceptualizes *healthy aging* as the process of developing and maintaining the functional ability that enables well-being in old age. One important domain of functional ability is to engage in social and cognitive activities in everyday life. In the literature, social activities refer to social participations and interpersonal interactions (Fratiglioni, Paillard-Borg, & Winblad, 2004). Cognitive activities include mentally stimulating activities (e.g., playing chess and playing a musical instrument), language use, remembering the past or imagining the future (Kemper & Sumner, 2001; Suddendorf & Corballis, 2007; Hertzog, Kramer, Wilson, & Lindenberger, 2008). Active participation in social and cognitive activities has been found to be associated with the maintenance of cognitive abilities and reduced risk of developing dementia in old age (Hertzog et al., 2008; Kuiper et al., 2015).

This chapter introduces the Electronically Activated Recorder (EAR), a naturalistic observation method, which is a portable audio recorder that periodically records sounds and speech in everyday life. We present how to use this method to objectively,

B. Demiray (✉) · M. Luo
Department of Psychology and University Research Priority Program "Dynamics of Healthy Aging", University of Zurich, Zurich, Switzerland
e-mail: b.demiray@psychologie.uzh.ch

A. Tejeda-Padron · M. R. Mehl
Department of Psychology, University of Arizona, Tucson, AZ, USA

P. L. Hill, M. Allemand (eds.), *Personality and Healthy Aging in Adulthood*, International Perspectives on Aging 26,
https://doi.org/10.1007/978-3-030-32053-9_8

reliably, and unobtrusively measure individual differences in everyday social and cognitive activities, and in turn, how these constructs can potentially associate with different personality and aging trajectories.

8.1 Assessing Individual Differences in Real-Life Social and Cognitive Activities

Social and cognitive activity participation has attracted much interest in aging research (e.g., Fratiglioni et al., 2004; Hertzog et al., 2008). Past studies have measured social and cognitive activity participation using the one-off self-report method: asking participants how often they have participated in an activity in a period of weeks, months, or years (Bielak, 2010; Salthouse, 2006). This method, despite its undisputed psychometric value and practical efficiency, can be subject to well-documented measurement limitations and biases, such as recall bias, response styles, demand characteristics, social desirability, and limitations to introspection (Schwarz, 2012). Moreover, the most commonly extracted metric from the self-report method is the average frequency of activity engagement over a period of time (Bielak, 2010). Using one aggregated score to represent a person's activity participation over time is one parsimonious way to measure *interindividual differences* in activity participation, i.e., differences between persons, in the population level. For example, the score of frequency of activity engagement, or even the score of variety of activity engagement (deriving from the count of yes/no endorsement of activities), has shown good associations with cognitive abilities in old age (Bielak, Mogle, & Sliwinski, 2019). However, interindividual differences do not only manifest in one single aggregated score, but also in the behavioral patterns that unfold over time, in a more micro level.

In fact, behaviors and psychological processes have been shown to vary across seconds, minutes, hours, days, and contexts (Houben, Van Den Noortgate, & Kuppens, 2015; Sherman, Rauthmann, Brown, Serfass, & Jones, 2015). In turn, *intraindividual variability* in activity participation, i.e., the fluctuation of activity within a person over the short time period and across contexts, can also contain information on how individuals differ from each other (Danvers, Wundrack, & Mehl, 2019; Paraschiv-Ionescu, Perruchoud, Buchser, & Aminian, 2012; Wang, Hamaker, & Bergeman, 2012; Ram et al., 2014). Ambulatory assessment techniques (Mehl & Conner, 2012), such as the diary method (e.g., Berntsen, 2007; Bielak, 2017; Bolger, Davis, & Rafaeli, 2003) and experience-sampling methods (Conner, Tennen, Fleeson, & Feldman Barrett, 2009; Heo, Lee, McCormick, & Pedersen, 2010) can be used to capture intraindividual variability in activity participation across short periods of time and across contexts, minimizing retrospective recall bias. However, these methods, as powerful as they are, are ultimately still based on self-report and can, thereby, be subject to limitations caused by the measurement burden imposed on the participants (e.g., having to interrupt one's normal activities multiple times per day for a few minutes to complete the prompted surveys), measurement-induced reactivity (e.g., the act of repeatedly introspecting on one's happiness; Conner & Reid, 2012), selection effects regarding which participants are able to sustain a motivation to self-report multiple times a day for several days (Scollon, Prieto, & Diener, 2009) and inability to remember, and therefore report, small, fleeting experiences and behaviors (e.g., Robbins, Karan, Lopez, & Weihs, 2018; Robbins, Mehl, Holleran, & Kasle, 2011).

In sum, in order to accurately capture information on social and cognitive engagements and to inform the understanding of health in old age, it is necessary to measure individuals' overall activity participation, as well as how they engage in social and cognitive activities throughout their everyday life and across time and contexts. As discussed above, particularly in measuring the latter type of information, it is not optimal to use traditional survey assessments or self-report based ambulatory assessments. A method that can reliably measure individuals' social and cognitive activities over time and across contexts, and that operates passively and that does not require participants' intensive engagement in data collection would be desirable. In this chapter, we review research with the EAR method that speaks to its potential to observationally, reliably, and unobtrusively measure social and cognitive activities in the real world, as well as environmental factors across time and contexts. Although, our focus in this chapter is specifically on the EAR method, it is important to note that the EAR is part of a broader set of observational/objective ambulatory assessment methods (Mehl & Conner, 2012) and, conceptually, is an "older sibling" to modern mobile sensing methods (Harari, Müller, Aung, & Rentfrow, 2017; Harari et al., 2016).

8.2 The EAR Method

The Electronically Activated Recorder (EAR) is an ambulatory assessment method for the unobtrusive observation of people's daily lives (Mehl, Pennebaker, Crow, Dabbs, & Price, 2001). Technically, it is a portable audio recorder that intermittently records ambient sound bites. Participants wear it while going about their days, unaware when exactly the device is recording. By tracking the ambient sounds in their lives, the EAR yields, in essence, an "acoustic diary" of their days. Through its sampling of brief sound bites, it protects privacy (i.e., takes snippets out of their larger conversational context) and enables larger scale empirical studies.

Originally, the EAR device was a modified, chip-triggered microcassette recorder. Harnessing the impressive development in mobile technologies over the last 20 years, it evolved, from the original analog audio recorder, first into a handheld digital audio recorder (akin to an mp3 recorder, the "digital EAR"), then into a recording software running on early pocket computers (the "Pocket EAR"), and finally into an app for commercial smartphones. After hosting the iEAR app for iOS devices (the "iEAR") for several years, Mehl and colleagues recently retired it and, instead, developed a fifth-generation EAR app that runs on Android smartphones (Mehl, 2017). In addition, researchers have begun to implement unobtrusive ambient audio sampling, which constitutes the heart of the EAR method, into other mobile devices, for example the uTrail, a small clip-on device that integrates audio sampling with GPS and accelerometry sampling (Röcke, Katana, Fillekes, Martin, & Weibel, 2018). Finally, unobtrusive audio sampling is also being integrated into modern mobile sensing methods (Harari et al., 2016, 2019).

In their research, Mehl and colleagues have found that wearing the EAR tends to be not very bothersome for participants and that participants usually demonstrate good compliance with wearing the EAR device. In a series of studies, the method has been successfully employed in age groups ranging from childhood to old age (3 years to 93 years; Alisic, Barrett, Bowles, Conroy, & Mehl, 2015; Bollich et al., 2016) and with different healthy and clinical populations (Baddeley, Pennebaker, & Beevers, 2013; Brown, Tragesser, Tomko, Mehl, & Trull, 2014). Over the last 20 years, more than two dozen investigator groups across several countries have employed the method and have, together, run more than 2000 participants through EAR monitoring protocols, suggesting the method's broad acceptance for studying a range of psychological questions (Alisic et al., 2015; Manson & Robbins, 2017; Minor, Davis, Marggraf, Luther, & Robbins, 2018; Tackman & Mehl, 2017).

The sampled raw ambient audio files are usually "converted" into quantitative psychological information (Tackman & Mehl, 2017). For that, trained research assistants listen to the EAR recordings, transcribe them and code them for objective behavioral aspects, such as participants' momentary activities (e.g., eating, watching TV, sleeping), emotional expressions (e.g., laughing, crying, sighing), interactions (e.g., dyadic conversation, group interaction, phone call), and locations (e.g., school, church, outdoors). In addition to these basic coding categories, researchers typically code study-specific categories that aim at capturing information that is relevant to their specific research questions (e.g., talking about cancer in research on how individuals cope with cancer, Robbins, López, Weihs, & Mehl, 2014; or talking about one's divorce experience in research on how individuals cope with a marital dissolution; Bourassa, Tackman, Mehl, & Sbarra, 2019). More detailed information on EAR coding is provided in Kaplan et al. (2019) and at the EAR Repository on the Open Science Framework (https://osf.io/n2ufd/).

With respect to the method's validity and psychometric utility, prior EAR research has yielded several findings that would be difficult to obtain with other methods, including traditional experience sampling. For example, Mehl and colleagues (2006) used the EAR to debunk the myth that women are more talkative than men. Analyzing data from almost 400 participants, they estimated that men and women both use about 16,000 words per day – with large interindividual differences, but no evidence of a sex difference. Similarly, Ramirez-Esparza, Mehl, Alvarez-Bermudez, and Pennebaker (2009) found, in a cross-cultural study, that American participants self-reported being more sociable than their Mexican counterparts, but Mexican participants actually spent less time alone, more time with others, and more time socializing. Further, when directly comparing self-perceived and EAR-assessed talkativeness, American participants perceived themselves as significantly more talkative than Mexicans, but Mexican participants spent approximately 9% more time talking.

Further, in a recent study, Mehl et al. (2017) found that information about daily word use derived from participants' daily interactions predicted biomarkers of inflammatory gene expression above and beyond standard self-report measures of negative emotionality (e.g., stress, anxiety, depression, loneliness). For example, the more spoken words the EAR captured for participants (as indicator of their objective level of social activity), the lower their level of inflammatory gene expression (as indicator of their body's preparedness to respond to stress with a pro-inflammatory immune response). Importantly, this effect of participants' daily verbosity, as an indicator of their objective level of social activity, proved independent of the effect of their reported loneliness, as an indicator of their subjective level of social activity.

As a final example, Sun and Vazire (2019) recently employed the EAR method to test people's self-knowledge of fluctuations in their personality states, that is to what degree people are aware of how extraverted, agreeable, conscientious and emotionally stable they really are on a moment-to-moment basis. For this, they compared participants' self-reported personality states against what the participants were "actually" like as determined via observer ratings of the recorded sound files. The findings showed that participants were relatively accurate in perceiving how extraverted and conscientious they were from moment to moment. Their perceived neuroticism and agreeableness states, however, evidenced little correspondence with their observed personality states. Sun and Vazire (2019) then argue that, for neuroticism, participants' subjective reports should be granted "truth value" (i.e. when participants said they were stressed, anxious, or depressed, that was really the case) and the observers just could not tell. For agreeableness, however, they argue that the EAR recordings should be taken as "ground truth" as they reflect how the participants really came across with others. And, there, participants lacked self-awareness in the sense that their subjective perceptions of how kind and compassionate they were did not match how others "objectively" heard them. For the study of self-knowledge, being able to measure subjective and objective (or inside and outside) perspectives independently, without shared method variance, is methodologically critical (Vazire, 2010).

Taken together, the psychometric utility of the EAR lies primarily in its ability to measure daily behavior unobtrusively, ecologically, and observationally, while minimizing shared method variance with global or momentary self-report. While most study applications of the EAR method so far have been in social, personality, health, and clinical psychology (Alisic et al., 2015; Allemand & Mehl, 2017; Mehl, 2017; Mehl & Wrzus, 2019; Wrzus & Mehl, 2015), researchers have recently begun to explore its potential for studying daily behavior and interaction processes in the context of (healthy) aging (e.g., Demiray, Mischler, & Martin, 2017).

8.3 Assessing Everyday Social Activities from Ecologically Sampled Ambient Sounds

For research on the interplay between personality and healthy aging, the EAR method can serve as a tool to study objective aspects of older adults' everyday social activities (e.g., to complement the study of subjective aspects using experience sampling). People's daily social activities have, on the one hand, strong theoretical (Wiggins, 1996) and empirical (Ashton, Lee, & Paunonen, 2002; Mehl, Gosling, & Pennebaker, 2006) relationships to their personalities At the same time, being able to maintain a meaningful, active social life well into old age is a robust predictor of aging well and longevity (e.g., Fratiglioni et al., 2004). Having few social network ties and infrequent social contact have been linked with detrimental health outcomes (e.g., Holt-Lunstad, Smith, Baker, Harris, & Stephenson, 2015). Low social connection is also a risk factor for incident dementia (Kuiper et al., 2015), poorer cognitive performance, faster cognitive decline and poorer executive functioning (e.g., Cacioppo & Hawkley, 2009).

Among other variables of potential interest (including the role of expressive behaviors such as sighing, Robbins et al., 2011, and swearing, Mascaro et al., 2018, and specific kinds of conversations such as gossip, Robbins & Karan, 2019), it is the quantity and quality of people's daily interactions that have been in the focus of prior EAR research. A key study linking these variables to well-being was published in 2010 by Mehl and colleagues (Mehl, Vazire, Holleran, & Clark, 2010). In this study, 79 college students wore the EAR for four days. The amount of time participants spent alone and the amount of time they spent talking with others were extracted as markers of the quantity of students' daily conversations. The amount of small talk (defined as an uninvolved, banal conversation where only trivial information was exchanged) and the amount of substantive conversations (defined as an involved conversation where meaningful information was exchanged) was extracted from the sampled ambient sounds. Across different measures of well-being (a) spending less time alone and more time interacting with others and (b) having less small talk and more substantive conversations were related to higher well-being.

Recognizing the small sample size and homogeneous population of this study, Milek et al. (2018) recently aimed at replicating this basic, but important finding. Using data from four EAR studies (with a combined sample size of almost 500 participants), they confirmed that (a) spending less time alone and more time interacting with others and (b) having more substantive conversations were again reliably related to higher levels of well-being. Interestingly, the original negative small talk effect failed to replicate suggesting that it may have been a false positive finding. Also interestingly, this study, due to its larger sample size, was able to test for the potential of personality to mediate (e.g., the association between interaction quantity and well-being being explained by extraversion) or moderate (e.g., the association between interaction quality and well-being being stronger among introverts) the effects. However, no significant mediation or moderation effect emerged, suggesting, for example, that introverts and extraverts tend to "run on the same conversational fuel". Having many daily interactions was not more closely linked to well-being for extraverts and having quality (rather than superficial) interactions was not a stronger predictor of well-being for introverts.

Together, these studies suggest that the association between the quality and quantity of people's daily social activity, as assessed via the EAR, and their level of well-being is a relatively broad and robust one. Therefore, these variables may constitute good candidate markers for future research on the role that daily social activities play in healthy aging. Specifically, since having strong social ties and an active social life style have been found to be able to buffer against cognitive decline and risk for dementia (Fratiglioni et al., 2004), the measurement of these factors, ideally passively and objectively in daily life, has high importance for understanding psychological processes involved in longevity and age-related quality of life (Ryan et al., 2019).

8.4 Assessing Everyday Cognitive Activities from Ecologically Sampled Ambient Sounds

As mentioned above, the main domain of application for the EAR method has been in social, personality and health psychology. Demiray and colleagues have adapted it for use in lifespan developmental psychology, with a focus on aging (e.g., Demiray et al., 2017; Luo, Robbins, Martin, & Demiray, 2019). Cognitive aging is perhaps one of the most important and popular domains of aging research. Cognitive aging theory and research, until recently, has been based on behavioral measures of cognitive performance assessed in the laboratory (e.g., reaction time, accuracy). Results from such studies have shown a general age-related decline in cognitive functions, such as episodic memory, working memory, attention and processing speed (e.g., Dennis & Cabeza, 2011). However, as proposed by the World Health Organization (2015), not only cognitive abilities, but also real-life cognitive activities are an essential part of understanding healthy aging. Furthermore, cognitive performance in laboratory tests does not always correspond to real-life cognitive activities. For example, this is evident in the widely observed age-prospective memory-paradox (e.g., Schnitzspahn, Ihle, Henry, Rendell, & Kliegel, 2011). This paradox is the general pattern of age-related deficits in prospective memory tasks conducted in the laboratory versus age-related benefits in prospective memory tasks that are carried out in participants' everyday lives. Therefore, it is extremely important to observe what older adults are actually doing in everyday life and to examine their cognitive activities across different real-life contexts.

Across a number of studies, Demiray and colleagues (e.g., Demiray, Mehl, & Martin, 2018; Demiray et al., 2017; Luo et al., 2019) have examined, for the first time, older adults' cognitive activities using the EAR method. They have focused mainly on memory-related activities (e.g., reminiscing about the personal past or imagining the future), and language use (e.g., vocabulary richness, grammar), as memory and language are a major part of cognitive aging research. They have conducted the first European EAR study (more detailed information on the Swiss EAR Study is provided on the Open Science Framework (https://osf.io/86fdr/) and coded participants' real-life conversations in terms of various cognitive activity indicators. For example, they manually coded for whether the participant was talking about their personal past (as an indicator of autobiographical memory/reminiscence) and personal future (as an indicator of prospection/episodic future thinking). Next, they automatically calculated scores for vocabulary richness and grammatical complexity as indicators of language use.

Remembering the Personal Past Reminiscence is the naturally occurring activity of thinking about or telling others about personally meaningful past experiences (Bluck & Levine, 1998). It often takes place spontaneously in daily life (in response to internal triggers such as thoughts or external triggers such as conversations, songs) or more deliberately in social contexts such as family meetings or class reunions (O'Rourke, King, & Cappeliez, 2017). Psychologists emphasize reminiscence and life review as a central task of old age and as promoting healthy aging (Erikson, 1959; Westerhof, Bohlmeijer, & Webster, 2010). Although the amount of reminiscence does not vary by age, young and old adults reminisce for different functions. For example, Pasupathi and Carstensen (2003) have shown that older adults experienced more positive emotions while reminiscing with others than young adults. The use of reminiscence in interventions and therapies for older adults is very common, emphasizing the relation between positive functions of reminiscence and psychological well-being (and between dysfunctional use of reminiscence and lower well-being; Pinquart & Forstmeier, 2012; Webster et al., 2010). Demiray et al. (2017) used the EAR method to observe social reminiscence in real life and explored how much, why and with whom older adults reminisce.

The sample in this study included 2164 sound files that included speech from 45 older adults. Reminiscence was coded in 110 of these files (5%), and reminiscence count ranged from 0 to 14 per participant with a mean of 2.44, which is 0–29.4% per participant. Thirteen participants never reminisced, which shows that 30% of the participants never engaged in social reminiscence during the four-day observation. These results suggest that there are large interindividual differences in the frequency of social reminiscence. Some older adults tend to be social "reminiscers" (Webster & Ma, 2013), whereas others do not engage in reminiscence at all in daily conversations. The observed interindividual differences can offer useful information for applied reminiscence research or in clinical settings, where reminiscence is used as a therapeutic tool (Pinquart & Forstmeier, 2012). Reminiscence interventions, especially in social settings (e.g., nursing homes), could be much more efficient with older adults who are naturally inclined to reminisce with others in everyday life. In contrast, older adults who are not interested in or used to reminiscing may be trained to reminisce in functional ways and to use reminiscence as a resource to enhance their well-being (O'Rourke, Cappeliez, & Claxton, 2011). The EAR method could be used to track older adults in everyday life before the intervention (pre-test), during intervention sessions, as well as post-intervention in order to evaluate the effectiveness of the intervention program.

The EAR method could also be used to investigate the content of reminiscence. When older adults are talking about their past in everyday life, what types of memories do they tend to retrieve and share with others? Furthermore, text analysis programs such as the Linguistic Inquiry and Word Count (LIWC; Pennebaker & Francis, 1999) could be used to explore "how" older adults talk about their important memories. For example, older adults who share more negative memories or who use a higher number of negative emotion words may be more likely to be depressed. Older adults who do not talk about specific one-time events or who do not give specific details about their memories may be more likely to show episodic memory deficits. Therefore, the EAR could be used as a diagnostic tool to detect preclinical cases and to prevent faster health decline.

Imagining the Personal Future Future time perspective is highly relevant for healthy aging, as aging, by definition, involves a narrowing future (e.g., Lang & Carstensen, 2002). Future time perspective represents an individual's perceptions of the future and his or her remaining time to live (e.g., Rohr, John, Fung, & Lang, 2017). Empirical evidence has emphasized the importance of this construct for shaping an individual's motivation, behavior and well-being (e.g., Cate & John, 2007; Kooij, Kanfer, Betts, & Rudolph, 2018). It is at the center of one of the most important aging theories, socioemotional selectivity theory, which posits that future time perspective shapes individuals' goals and social relations across the adult lifespan (e.g., Carstensen, Isaacowitz, & Charles, 1999; Lang & Carstensen, 2002). Past research shows that older adults who are able to maintain a positive and open-ended future time perspective have higher levels of well-being (e.g., Demiray & Bluck, 2014). Future time perspective is measured mostly via self-report, specifically with the Future Time Perspective Scale (FTPS; Carstensen & Lang, 1996). Brianza and Demiray (2019) have examined future time perspective, for the first time, via both a subjective measure (FTPS) and two objective EAR measures (how much and how people talk about their future in everyday life). They have counted participants' real-life utterances about their future, and the types of words they used while talking about their future, and examined whether these two behaviors could be used as objective indicators of their future time perspective. They have explored the relation of these behaviors to the self-report measure of FTPS. That is, they explored whether individuals' subjective and global perception of their future was associated with how much and how they talked about their future. Finally, they examined the widely studied relation between future time perspective and life satisfaction.

Their results showed that there was no relation between FTPS scores and how much and how individuals talked about their future. Furthermore, this result held for both young and older adults. Although talking behavior was unrelated to subjective future time perspective, it was still associated with life satisfaction: Young adults' life satisfaction was predicted by both their subjective perception of the future, and by how much they talked about the future and used family-related words. For older adults, the most important predictor of life satisfaction was their subjective view of the future, followed by how much they uttered achievement-related words. That is, for older adults, it is not how much they *talk* about their future, but how they *think* about their future that counts. Young adults may be creating their futures both privately and socially (in conversations), whereas older adults tend to talk less about their future, which has no impact on their life satisfaction. This suggests that self-report may be a more suitable method to assess older adults' future time perspective, whereas young adults' future time perspective may be more diversely represented in different behaviors and suitable to be measured in alternative ways. It seems that momentary self-report (e.g., experience-sampling) might be a more suitable method to examine older adults' future time perspective in everyday life and its intraindividual variations across contexts (Allemand & Hill, 2019).

Conversational Time Travel Recently, psychologists have started to emphasize that recalling the personal past and imagining the personal future are closely related phenomena that share many common qualities (e.g., Schacter et al., 2012; Klein, 2013). Mental time travel refers to this ability to recall experiences from the past and to imagine possible events in the future (Suddendorf & Corballis, 2007). Aging adds a new dimension to the experience of mental time travel: Older adults have a long past full of memories with a limited future time perspective, whereas younger individuals have narrower life stories with a more open-ended future time perspective (e.g., Demiray & Bluck, 2014). This should influence the nature and functions of their mental time travel and how it relates to their well-being.

In applying the concept of mental time travel to real-life conversations, Demiray, Mehl and Martin (2018) developed the term *conversational time travel*. They explored, for the first time, how much young and older individuals talk about their personal past versus future. There is a prospective bias observed in mental time travel (people tend to *think* more about their personal future than their past; e.g., Felsman, Verduyn, Ayduk, & Kross, 2017; Song & Wang, 2012). The authors explored whether this holds for conversational time travel and whether it holds across different samples (American vs. Swiss, young vs. old, breast cancer patients vs. healthy adults). They developed a new coding scheme for conversational time travel and found that participants talked about their personal past two to three times as much as their personal future. This finding was consistent across young adults, healthy older adults, breast cancer patients and healthy spouses of the patients (past: 10.1–13.6%, future: 2.7–7.2%). This suggests that the retrospective bias in conversational time travel (i.e., participants *talk* more about the past than future) might be a universal phenomenon independent of age, and it is in contrast with the prospective bias in mental time travel. Although older adults have a larger past and a narrower future compared to the young, they still talk about their past and future as much as young do. Older adults do not seem to live in the past (at least in conversations), as suggested by aging stereotypes.

Although this retrospective bias seems to be a robust finding, there were still large interindividual differences in how much people talked about their personal past versus future. This might be related to individuals' trait-like temporal orientation, which refers to relatively stable interindividual differences in the relative emphasis one places on the past, present, or future (Zimbardo & Boyd, 1999). Zimbardo and colleagues have differentiated between distinct temporal orientations (e.g., "past-negative" type), which have been widely examined in relation to personality traits (e.g., Zhang & Howell, 2011) and health outcomes (see Stolarski, Fieulaine, & Beek, 2015 for reviews). These trait-like tendencies might reflect onto individuals' utterances in real life and can be tracked with the EAR method. For example, future research can explore whether an aging individual who has been categorized as a "past-negative" type is more likely to talk about the past (than future) and more likely to talk about the past in a negative way (than a positive way) in real life.

Demiray, Luo and Martin (in preparation) built on the work on conversational time travel and explained why young and older individuals talked more about their past than their future: They examined intraindividual differences in conversational time travel across its different functions. Past questionnaire data show that there are three major functions of mental time travel: self, social and directive. Individuals self-report thinking or talking about their personal past and/or future in order to maintain and enhance their sense of self, to socialize with others, and to guide and direct their behavior (e.g., Bluck, Alea, & Demiray, 2010; Harris, Rasmussen, & Berntsen, 2014). The authors objectively coded past- and future-oriented utterances in terms of these functions and found that young and older adults talked about their future more for directive purposes (i.e., learning, planning, decision making, problem solving), whereas they talked about their past more with social functions (i.e., making conversation, teaching, giving advice). This suggests that past-oriented conversations are more useful or functional in social interactions (compared to future-oriented ones) and, thus, people talk more about their past in social interactions (Demiray et al., 2018).

The authors also examined the relation between conversational time travel and laughing behavior (as an indicator of good mood). They found that both young and older adults laughed more while talking about their past than their future. In fact, participants were less likely to laugh, while talking about their future. The study showed that the association between laughing and conversational time travel was not associated with the interindividual differences of age: The association was the same across young and healthy older adults. In conclusion, people tend to share their laughters with others, and it seems that both young and older adults use their past as a resource to laugh and feel good about themselves while socializing with others. These results should be informative in clinical settings for the promotion of healthy aging, as well as the prevention of depression or dementia in the elderly while using reminiscence therapies.

Language Use Language use in everyday life (e.g., vocabulary richness, grammatical complexity) is another cognitive activity that can be efficiently assessed via the EAR. Past studies that examined the effects of cognitive aging on language use have relied on speech samples from speech production tasks in the laboratory or from telephone conversations with strangers (e.g., Horton, Spieler, & Shriberg, 2010; Kemper & Sumner, 2001). These speech samples may not be representative of language use in naturally occurring conversations. Luo and colleagues contribute to the literature by using the EAR to observe language use in natural contexts in the real world (e.g., Luo et al., 2019).

Using the data from the Swiss EAR study, Luo, Schneider, Martin, and Demiray (2019) examined, for the first time, age effects in vocabulary richness and grammatical complexity across two types of social contexts that have been shown relevant to language use: activities (i.e., socializing and working) and conversation types (i.e., small talk and substantive conversation; Levinson, 1992). Multilevel models showed that vocabulary richness and grammatical complexity increased during socializing and substantive conversations, but decreased in small talk. Furthermore, older adults used richer vocabulary and more complex grammatical structures at work than young adults. They also used richer vocabulary in small talk. In contrast, young adults used richer vocabulary than older adults during non-socializing and non-working occasions, such as watching TV and exercising. That is, intraindividual differences in language use across different social contexts were associated with interindividual differences (i.e., age). This paper shows the value of examining real-life activities in combination with contextual information for healthy aging research. Furthermore, Mehl et al. (2006) have shown that personality is associated with language use in everyday life. For example, conscientiousness was negatively related to the use of swear words. Future studies can examine whether personality is associated with vocabulary richness and grammatical complexity in real-life language use.

In sum, the EAR method enables researchers to objectively and unobtrusively measure cognitive activity participation in everyday life. From the massive amount of audio files, we have been able to extract information on interindividual differences and intraindividual variability in memory (e.g., conversational time travel) and language-related activities across different contexts. These observations in behaviors and activities can be linked with interindividual differences in personality and health in old age. Moreover, we have also observed, across studies, a lack of associations between real-life activity participation and global traits or abilities measured in the laboratory. For example, there was no correlation between participants' real-life vocabulary richness and their maximum verbal ability measured by vocabulary test performance in the laboratory (Luo et al., 2019). There was no association between the frequency or qualities of future-oriented utterances in real life and global future time perspective (Brianza & Demiray, 2019). Similarly, Wank et al. (2019) showed that real-life episodic autobiographical memory specificity was very weakly related to a neuropsychological measure of episodic memory. These results indicate that the pure focus on interindividual differences in capacity (for example, in vocabulary test scores) may be insufficient to understand what aging individuals are actually doing in real life. In contrast, the combination of interindividual differences (such as age) and contextual information is useful in understanding real-life activities (Luo et al., 2019). Thus, adding contextual information can enhance the understanding of how older individuals vary their cognitive activities from context to context. The contextual information is readily available in the EAR sound files and can be easily extracted.

Ongoing and Future EAR Research on Healthy Aging As reviewed above, the EAR method has been very helpful in objectively observing aging individuals' social and cognitive activities in the real world. Thus, we aim to expand our repertoire of real-life activities with novel EAR studies that build on previous work with more sophisticated study designs, more advanced technology use, and research questions that can test the WHO healthy aging model. Individuals' personalities are reflected onto their thoughts, feelings and behaviors in everyday life, and the EAR can capture these aspects to the extent that they are observable. One advancement in research methodology will be to embrace a multi-method approach in which the EAR is supported with other ambulatory assessment techniques, such as mobile sensing (to collect additional sensory data, such as GPS) and momentary self-report (to gather information on unshared thoughts and feelings that cannot be detected by the EAR).

A novel EAR study conducted in Switzerland is the Mobility, Activity and Social Interactions Study (MOASIS) on healthy older adults (Röcke, Katana, Fillekes, Martin, & Weibel, 2018). The MOASIS Project used a custom-built mobile sensor, the uTrail, to collect GPS and accelerometer data, as well as EAR data, to collect objective information on the spatial and physical activities of older adults, in addition to social and cognitive activities. Participants carried the uTrail with them for 30 days, which was programmed to record 50 seconds of ambient sounds every 18 minutes with a black-out period at night.

In this project, participants were also provided with a smartphone and asked to carry it (in addition to the uTrail) for 2 weeks to complete an experience-sampling survey eight times a day. Thus, EAR recordings were complemented with the experience-sampling of momentary thoughts, feelings, and activities (which might reflect personality states), and cognitive performance in an ambulatory working memory task. One limitation of the study is that the sound recordings and experience-sampling surveys were not aligned and occurred independently of each other. Thus, our first task is to detect the sound recordings that occurred within a very short amount of time (e.g., 5 min) with the surveys (e.g., Sun, Schwartz, Son, Kern, & Vazire, 2019). These objective (EAR) and subjective (self-report) sources of data will be combined to create a stronger multi-method approach to studying personality in older adults as reflected in their momentary thoughts, feelings, conversations and activities. For example, an older adult who rated their mood as very positive in a given experience-sampling survey may have been in a pleasant conversation with many laughters in the corresponding sound file (Ramírez-Esparza et al., 2019). Future research should time-align sound recordings with experience-sampling surveys (e.g., sound recording occurring right before a signal is sent to the participant) to optimize this goal. Finally, audio data should be combined with other sources of objective data from participants' lives (i.e., GPS, accelerometer) to answer novel questions about social or cognitive activities, such as "On days when older adults are physically more active and diverse – as measured by GPS and accelerometer –

are they also more active and diverse in their social interactions, and more diverse and complex in their language use?". In sum, the harmonization of different sources of ambulatory assessment data bears strong potential for improving our understanding of interindividual differences and intraindividual variations in aging-related activities.

Another study that used the EAR is the RHYTHM (Realizing Healthy Years Through Health Maintenance) Study (e.g., Aschwanden, Luchetti, & Allemand, 2019). RHYTHM was designed to examine how older individuals actively use multiple health stabilization processes and maintenance activities in everyday life. Participants carried a smartphone with them for 10 days, which was programmed to record 30 seconds of ambient sounds every 12.5 minutes between 8 am and 8 pm. Similar to the MOASIS Project, EAR recordings were complemented with the experience-sampling of momentary thoughts, affect, activities and so on, and were aligned with experience-sampling surveys within a 2-hour time window (e.g., Sun et al., 2019). One major research question in this project is how self-reported momentary personality states might be reflected onto actual social and cognitive activities detected by the EAR. For example, an older adult who has reported feeling highly extraverted within a given 2-hour time window may have engaged in many conversations with different people as detected by the EAR.

In sum, in line with the WHO healthy aging model, these two new studies will obtain rich within-person profiles of older adults' daily life activities, performance, experiences and the context in which these occur. This suggests that the EAR has become a highly popular and efficient method to observe the everyday activities of older adults.

Recommendations and Drawbacks Regarding the EAR The EAR method allows us to collect hundreds of thousands of audio files from individuals' everyday lives. However, listening to all audio files and manually transcribing and coding them is a very laborious and demanding task. It takes years of hard work and a large, well-trained team of coders to reliably extract meaningful psychological information from the sound files (for a review of best practices for EAR data coding and processing, see Kaplan et al., 2019). Thus, social scientists using the EAR method should and have started to invest in the automatization of these coding procedures (Dubey, Mehl, & Mankodiya, 2016).

For example, Demiray and colleagues have collaborated with computer scientists to develop a computer program that automatically estimates the presence (or absence) of human voice in real-life audio data (Cheetham, Demiray, Martin, Battegay & Londral, under review). Similar software tools are available (e.g., Asgari, Shafran, & Bayestehtashk, 2012; Harari et al., 2019; Wyatt, Choudhury, Bilmes, & Kitts, 2011), but are often inaccessible to researchers with no or little programming background. The current program has a user-friendly interface and estimates for every sound file whether it is a) highly likely to contain human voice, b) not likely to contain human voice or c) the algorithm cannot tell with certainty. To the extent that the algorithm evidences high accuracy across a range of populations, contexts, and languages, it can enable researchers to save a significant amount of time. As more than half of EAR sound files do not include participant speech (e.g., Demiray et al., 2018; Milek et al., 2018; Robbins, López, Weihs, & Mehl, 2014), researchers who are only (or mainly) interested in social interactions and conversations can automatically ignore sound files that do not include human voice.

One limitation of the program currently is that it can classify TV and radio as human voice as well (to the extent that they contain speech), which are misread as "social interaction". For example, an older adult who watches TV by himself all day could be mistaken for a "very social personality" based on the program. Similarly, certain harmonic sounds (e.g., the sound of a fan or a blender) have a potential to be misclassified as speech (Wyatt et al., 2011). In practice, then, the initial automatic classification of sound files as containing human voice or not requires a follow-up human review of the classification. The current program, indeed, allows researchers to manually inspect each sound file and its automatic classification to validate the result. Practically, it is helpful that the program's rate of false negative decisions (i.e., human voice misclassified as no voice) tends to be quite low relative to the rate of false positive decisions (i.e., no voice misclassified as voice). Future research is necessary to improve automatic conversation detection algorithms, for example, through incorporation of foreground and background speech recognition (Feng, Nadarajan, Vaz, Booth, & Narayanan, 2018; Nadarajan, Somandepalli, & Narayanan, 2019) and implementation of speaker identification (e.g., to automatically differentiate the participant's voice from bystanders' voices).

In another line of research, Demiray, Mehl and colleagues are collaborating with computer scientists to use machine learning to automatically code the transcripts of real-life speech data. For example, Yordanova, Demiray, Mehl and Martin (2019) have trained a classifier to code for the participant's conversation partner, location, activity, mood, temporal focus of conversation (e.g., past, future) and functions of conversation, and found that the classifier was able to code transcripts with 74–99% accuracy. For example, 74% accuracy was reached for coding whether the function of the conversation was identity-related or not. 99% accuracy was achieved for coding whether the participant was talking about aging or not. These high accuracy rates are a significant advantage for social scientists who work with "big" transcribed datasets, which will save substantial amounts of time and energy spent on manual coding. However, this approach still requires verbatim transcripts, which requires human transcribers. The reason is that it is extremely difficult to achieve sufficient accuracy in automatic speech-to-text transcription in real-life environments, where ambient noise is high (D. Imseng, personal communication, May 17, 2018).

Finally, a novel and active area of research focuses on the development of automated, computerized methods to detect psychological information through acoustic analyses. That is, researchers are trying to eliminate all transcription and coding procedures by directly extracting meaningful information from sound or voice features (e.g., intensity, pitch, energy). For example, acoustic features are automatically analyzed to detect the emotions of the speaker (e.g., Schmitt, Ringeval, & Schuller, 2016). Recently, Weidman et al. (2019) focused on one emotional state (i.e., happy mood) and asked participants to self-report their state happy mood, as well as providing audio recordings in everyday life. They examined whether acoustic analysis could automatically detect fluctuations in happy mood. They used three different machine learning algorithms, but found that they showed minimal predictive power above chance. They concluded that it is not, yet, possible to automatically detect intraindividual variations in one emotional state (i.e., happy mood). This suggests that rigorous research is necessary in this new field in order to completely automatize the processing of audio data.

8.5 Conclusions

Social and cognitive engagements are important domains of functional ability in the healthy aging model of the World Health Organization (2015) and have been shown to include important information about health in old age. The prior endeavors in measuring and understanding everyday social and cognitive activities will be continued. The EAR method enables researchers to unobtrusively and reliably gather information on real-life social and cognitive activity participation, such as engaging in small talk and substantive conversations, reminiscing about the past versus imagining the future, and producing language. We have discussed how the interindividual differences in these social and cognitive activities can be associated with personality and health in old age. We have also discussed measuring intraindividual variability in these activities over time and across contexts, and how they could be associated with different personality and health trajectories. For example, the intraindividual variability in vocabulary richness and grammatical complexity varied across different social contexts and were associated with different ages. In sum, the EAR method offers reliable real-life evidence for the understanding of activity participation in the context of healthy aging.

Furthermore, the EAR method can involve older adults in research who would otherwise be excluded from real-life studies relying on self-report. For example, older adults who are intimidated by technology, unable to use a smartphone to complete experience-sampling surveys, or unable to self-report due to worsened eyesight could be tracked in real life with the EAR. Clinical populations such as depression, dementia or Alzheimer's disease patients, as well as patients with multimorbidity could also be naturally observed via the EAR. Finally, this method would be suitable for cross-cultural studies, as well as for studies focusing on diversity (e.g., older adults with low SES, low education level, in rural vs. urban areas), as the EAR does not require any self-report from the participants. All that is needed from the participant is that they carry the EAR with them in everyday life (or have the EAR in vicinity in case of patients) in order to track their social and cognitive activities.

References

Alisic, E., Barrett, A., Bowles, P., Conroy, R., & Mehl, M. R. (2015). Topical review: Families coping with child trauma: A naturalistic observation methodology. *Journal of Pediatric Psychology, 41*, 117–127.

Allemand, M., & Hill, P. L. (2019). Future time perspective and gratitude in daily life: A micro-longitudinal study. *European Journal of Personality, 33*(3), 385–399.

Allemand, M., & Mehl, M. R. (2017). Personality assessment in daily life: A roadmap for future personality development research. In J. Specht (Ed.), *Personality development across the lifespan* (pp. 437–454). San Diego, CA: Elsevier.

Aschwanden, D., Luchetti, M., & Allemand, M. (2019). Are open and neurotic behaviors related to cognitive behaviors in daily life of older adults? *Journal of Personality, 87*, 472–484.

Asgari, M., Shafran, I., & Bayestehtashk, A. (2012). Robust detection of voiced segments in samples of everyday conversations using unsupervised hmms. In *2012 IEEE spoken language technology workshop (SLT)* (pp. 438–442). Berkeley, CA: IEEE.

Ashton, M. C., Lee, K., & Paunonen, S. V. (2002). What is the central feature of extraversion? Social attention versus reward sensitivity. *Journal of Personality and Social Psychology, 83*, 245–252.

Baddeley, J. L., Pennebaker, J. W., & Beevers, C. G. (2013). Everyday social behavior during a major depressive episode. *Social Psychological and Personality Science, 4*, 445–452.

Berntsen, D. (2007). Involuntary autobiographical memories: Speculations, findings, and an attempt to integrate them. In J. H. Mace (Ed.), *New perspectives in cognitive psychology. Involuntary memory* (pp. 20–49). Malden, MA: Blackwell Publishing.

Bielak, A. A. (2010). How can we not 'lose it' if we still don't understand how to 'use it'? Unanswered questions about the influence of activity participation on cognitive performance in older age–a mini-review. *Gerontology, 56*, 507–519.

Bielak, A. A. (2017). Different perspectives on measuring lifestyle engagement: A comparison of activity measures and their relation with cognitive performance in older adults. *Aging, Neuropsychology, and Cognition, 24*, 435–452.

Bielak, A. A., Mogle, J. A., & Sliwinski, M. J. (2019). Two sides of the same coin? Association of variety and frequency of activity with cognition. *Psychology and Aging, 34*, 457–466.

Bluck, S., Alea, N., & Demiray, B. (2010). You get what you need: The psychosocial functions of remembering. In J. H. Mace (Ed.), *New perspectives in cognitive psychology. The act of remembering: Toward an understanding of how we recall the past* (pp. 284–307). Hoboken, NJ: Wiley-Blackwell.

Bluck, S., & Levine, L. J. (1998). Reminiscence as autobiographical memory: A catalyst for reminiscence theory development. *Ageing and Society, 18*, 185–208.

Bolger, N., Davis, A., & Rafaeli, E. (2003). Diary methods: Capturing life as it is lived. *Annual Review of Psychology, 54*, 579–616.

Bollich, K. L., Doris, J. M., Vazire, S., Raison, C. L., Jackson, J. J., & Mehl, M. R. (2016). Eavesdropping on character: Assessing everyday moral behaviors. *Journal of Research in Personality, 61*, 15–21.

Bourassa, K. J., Tackman, A. M., Mehl, M. R., & Sbarra, D. A. (2019). Psychological overinvolvement, emotional distress, and daily affect following marital dissolution. *Collabra: Psychology, 5*(1), 8.

Brianza, E., & Demiray, B. (2019). Future time perspective and real-life utterances about the future in young and older adults. *The Journal of Gerontopsychology and Geriatric Psychiatry.*

Brown, W. C., Tragesser, S. L., Tomko, R. L., Mehl, M. R., & Trull, T. J. (2014). Recall of expressed affect during naturalistically observed interpersonal events in those with borderline personality disorder or depressive disorder. *Assessment, 21*, 73–81.

Cacioppo, J. T., & Hawkley, L. C. (2009). Perceived social isolation and cognition. *Trends in Cognitive Sciences, 13*, 447–454.

Carstensen, L. L., Isaacowitz, D. M., & Charles, S. T. (1999). Taking time seriously: A theory of socioemotional selectivity. *American Psychologist, 54*, 165–181.
Carstensen, L. L., & Lang, F. R. (1996). *Future time perspective scale*. Unpublished manuscript, Stanford University.
Cate, R. A., & John, O. P. (2007). Testing models of the structure and development of future time perspective: Maintaining a focus on opportunities in middle age. *Psychology and Aging, 22*, 186–201.
Cheetham, M., Demiray, B., Martin, M., Battegay, E., & Londral, A. (under review). Semi-automatic human voice detection and segmentation: A method for use in behavioural speech analysis
Conner, T. S., & Reid, K. A. (2012). Effects of intensive mobile happiness reporting in daily life. *Social Psychological and Personality Science, 3*, 315–323.
Conner, T. S., Tennen, H., Fleeson, W., & Feldman Barrett, L. (2009). Experience sampling methods: A modern idiographic approach to personality research. *Social and Personality Psychology Compass, 3*, 292–313.
Danvers, A. F., Wundrack, R., & Mehl, M. R. (2019). *Equilibria in personality states*. Manuscript under review.
Demiray, B., & Bluck, S. (2014). Time since birth and time left to live: Opposing forces in constructing psychological wellbeing. *Ageing and Society, 34*, 1193–1218.
Demiray, B., Luo, M., & Martin, M. (in preparation). People laugh more while talking about their past than their future: Functions of conversational time travel in everyday life.
Demiray, B., Mehl, M. R., & Martin, M. (2018). Conversational time travel: Evidence of a retrospective bias in real life conversations. *Frontiers in Psychology, 9*, Article 2160.
Demiray, B., Mischler, M., & Martin, M. (2017). Reminiscence in everyday conversations: A naturalistic observation study of older adults. *The Journals of Gerontology: Series B, 74*, 745–755.
Dennis, N. A., & Cabeza, R. (2011). Neuroimaging of healthy cognitive aging. In F. I. M. Craik & T. A. Salthouse (Eds.), *The handbook of aging and cognition* (pp. 10–63). New York: Psychology Press.
Dubey, H., Mehl, M. R., & Mankodiya, K. (2016). BigEAR: Inferring the ambient and emotional correlates from smartphone-based acoustic Big Data. *IEEE First International Conference on Connected Health: Applications, Systems and Engineering Technologies (CHASE)*, Washington, DC, USA.
Erikson, E. H. (1959). Identity and the life cycle: Selected papers. *Psychological Issues, 1*, 1–171.
Felsman, P., Verduyn, P., Ayduk, O., & Kross, E. (2017). Being present: Focusing on the present predicts improvements in life satisfaction but not happiness. *Emotion, 17*, 1047–1051.
Feng, T., Nadarajan, A., Vaz, C., Booth, B., & Narayanan, S. (2018). TILES audio recorder: An unobtrusive wearable solution to track audio activity. In *Proceedings of the fourth ACM workshop on wearable systems and applications* (pp. 33–38). New York, NY: ACM.
Fratiglioni, L., Paillard-Borg, S., & Winblad, B. (2004). An active and socially integrated lifestyle in late life might protect against dementia. *The Lancet Neurology, 3*, 343–353.
Harari, G. M., Lane, N. D., Wang, R., Crosier, B. S., Campbell, A. T., & Gosling, S. D. (2016). Using smartphones to collect behavioral data in psychological science: Opportunities, practical considerations, and challenges. *Perspectives on Psychological Science, 11*, 838–854.
Harari, G. M., Müller, S. R., Aung, M. S., & Rentfrow, P. J. (2017). Smartphone sensing methods for studying behavior in everyday life. *Current Opinion in Behavioral Sciences, 18*, 83–90.
Harari, G. M., Müller, S. R., Stachl, C., Wang, R., Wang, W., Bühner, M.,… Gosling, S. D. (2019). Sensing sociability: Individual differences in young adults' conversation, calling, texting, and app use behaviors in daily life. *Journal of Personality and Social Psychology.*
Harris, C. B., Rasmussen, A. S., & Berntsen, D. (2014). The functions of autobiographical memory: An integrative approach. *Memory, 22*, 559–581.
Heo, J., Lee, Y., McCormick, B. P., & Pedersen, P. M. (2010). Daily experience of serious leisure, flow and subjective well-being of older adults. *Leisure Studies, 29*, 207–225.

Hertzog, C., Kramer, A. F., Wilson, R. S., & Lindenberger, U. (2008). Enrichment effects on adult cognitive development: Can the functional capacity of older adults be preserved and enhanced? *Psychological Science in the Public Interest, 9*, 1–65.

Holt-Lunstad, J., Smith, T. B., Baker, M., Harris, T., & Stephenson, D. (2015). Loneliness. and social isolation as risk factors for mortality: A meta-analytic review. *Perspectives on Psychological Science, 10*, 227–237.

Horton, W. S., Spieler, D. H., & Shriberg, E. (2010). A corpus analysis of patterns of age-related change in conversational speech. *Psychology and Aging, 25*, 708–713.

Houben, M., Van Den Noortgate, W., & Kuppens, P. (2015). The relation between short-term emotion dynamics and psychological well-being: A meta-analysis. *Psychological Bulletin, 141*, 901–930.

Kaplan, D. M., Rentscher, K. E., Lim, M., Keating, D., Romero, J., Shah, A., …, & Mehl, M. R. (2019). Best practices for electronically activated recorder (EAR) data coding and processing: A practical guide for researchers. *Behavioral Research Methods*. https://doi.org/10.31234/osf.io/mgzcy

Kemper, S., & Sumner, A. (2001). The structure of verbal abilities in young and older adults. *Psychology and Aging, 16*, 312–322.

Klein, W. (2013). *Time in language*. New York, NY: Routledge.

Kooij, D., Kanfer, R., Betts, M., & Rudolph, C. W. (2018). Future time perspective: A systematic review and meta-analysis. *Journal of Applied Psychology, 103*, 867–893.

Kuiper, J. S., Zuidersma, M., Voshaar, R. C. O., Zuidema, S. U., van den Heuvel, E. R., Stolk, R. P., & Smidt, N. (2015). Social relationships and risk of dementia: A systematic review and meta-analysis of longitudinal cohort studies. *Ageing Research Reviews, 22*, 39–57.

Lang, F. R., & Carstensen, L. L. (2002). Time counts: Future time perspective, goals, and social relationships. *Psychology and Aging, 17*, 125–139.

Levinson, S. C. (1992). Activity types and language. In P. Drew & J. Heritage (Eds.), *Talk at work: Interaction in institutional settings* (pp. 66–100). New York, NY: Cambridge University Press.

Luo, M., Robbins, M., Martin, M., & Demiray, B. (2019a). Real-life language use across different interlocutors: A naturalistic observation study of adults varying in age. *Frontiers in Psychology, 10*, 1412.

Luo, M., Schneider, G., Martin, M., & Demiray, B. (2019b). Cognitive aging effects on language use in real-life contexts: a naturalistic observation study. *Proceedings of the 41th Annual Meeting of the Cognitive Science Society*.

Manson, J. H., & Robbins, M. L. (2017). New evaluation of the electronically activated recorder (ear): Obtrusiveness, compliance, and participant self-selection effects. *Frontiers in Psychology, 8*, 658.

Mascaro, J. S., Rentscher, K. E., Hackett, P. D., Lori, A., Darcher, A., Rilling, J. K., & Mehl, M. R. (2018). Preliminary evidence that androgen signaling is correlated with men's everyday language. *American Journal of Human Biology, 30*, e2316.

Mehl, M. R. (2017). The electronically activated recorder (EAR): A method for the naturalistic observation of daily social behavior. *Current Directions in Psychological Science, 26*, 184–190.

Mehl, M. R., & Conner, T. S. (2012). *Handbook of research methods for studying daily life*. New York: Guilford Press.

Mehl, M. R., Gosling, S. D., & Pennebaker, J. W. (2006). Personality in its natural habitat: Manifestations and implicit folk theories of personality in daily life. *Journal of Personality and Social Psychology, 90*, 862–877.

Mehl, M. R., Pennebaker, J. W., Crow, D. M., Dabbs, J., & Price, J. H. (2001). The electronically activated recorder (EAR): A device for sampling naturalistic daily activities and conversations. *Behavior Research Methods, 33*, 517–523.

Mehl, M. R., Raison, C. L., Pace, T. W. W., Arevalo, J. M. G., & Cole, S. W. (2017). Natural language indicators of differential gene regulation in the human immune system. *PNAS, 114*, 12554–12559.

Mehl, M. R., Vazire, S., Holleran, S. E., & Clark, C. S. (2010). Eaves-dropping on happiness: Well-being is related to having less small talk and more substantive conversations. *Psychological Science, 21*, 539–541.

Mehl, M. R., & Wrzus, C. (2019). Ecological sampling methods for studying personality in daily life. In O. P. John & R. W. Robins (Eds.), *The handbook of personality* (4th ed.). New York, NY: Guilford Press.

Milek, A., Butler, E. A., Tackman, A. M., Kaplan, D. M., Raison, C. L., Sbarra, D. A., … Mehl, M. R. (2018). "Eavesdropping on happiness" revisited: A pooled, multisample replication of the association between life satisfaction and observed daily conversation quantity and quality. *Psychological Science, 29*, 1451–1462.

Minor, K. S., Davis, B. J., Marggraf, M. P., Luther, L., & Robbins, M. L. (2018). Words matter: Implementing the electronically activated recorder in schizotypy. *Personality Disorders, Theory, Research, and Treatment, 9*, 133–143.

Nadarajan, A., Somandepalli, K., & Narayanan, S. S. (2019). Speaker agnostic foreground speech detection from audio recordings in workplace settings from wearable recorders. In *ICASSP 2019–2019 IEEE International Conference on Acoustics, Speech and Signal Processing (ICASSP)* (pp. 6765–6769). Piscataway, NJ: IEEE.

O'Rourke, N., Cappeliez, P., & Claxton, A. (2011). Functions of reminiscence and the psychological well-being of young-old and older adults over time. *Aging & Mental Health, 15*, 272–281.

O'Rourke, N., King, D. B., & Cappeliez, P. (2017). Reminiscence functions over time: consistency of self-functions and variation of prosocial functions. *Memory, 25*, 403–411.

Paraschiv-Ionescu, A., Perruchoud, C., Buchser, E., & Aminian, K. (2012). Barcoding human physical activity to assess chronic pain conditions. *PLoS One, 7*, e32239.

Pasupathi, M., & Carstensen, L. L. (2003). Age and emotional experience during mutual reminiscing. *Psychology and Aging, 18*(3), 430.

Pennebaker, J., & Francis, M. (1999). Linguistic inquiry and word count: LIWC, 1999.

Pinquart, M., & Forstmeier, S. (2012). Effects of reminiscence interventions on psychosocial outcomes: A meta-analysis. *Aging & Mental Health, 16*(5), 541–558.

Ram, N., Conroy, D. E., Pincus, A. L., Lorek, A., Rebar, A., Roche, M. J., … Gerstorf, D. (2014). Examining the interplay of processes across multiple time-scales: Illustration with the Intraindividual Study of Affect, Health, and Interpersonal Behavior (iSAHIB). *Research in Human Development, 11*, 142–160.

Ramírez-Esparza, N., Garcia-Sierra, A., Rodriguez-Arauz, G., Ikizer, E. G., & Fernández-Gómez, M. J. (2019). No laughing matter: Latinas' high quality of conversations relate to behavioral laughter. *PLoS One, 14*, e0214117.

Ramírez-Esparza, N., Mehl, M. R., Álvarez-Bermúdez, J., & Pennebaker, J. W. (2009). Are Mexicans more or less sociable than Americans? Insights from a naturalistic observation study. *Journal of Research in Personality, 43*, 1–7.

Robbins, M. L., & Karan, A. (2019). Who gossips and how in everyday life. *Social Psychological and Personality Science*. https://doi.org/10.1177/1948550619837000

Robbins, M. L., Karan, A., Lopez, A. M., & Weihs, K. (2018). Naturalistically observing non-cancer conversations among couples coping with breast cancer. *Psycho-Oncology, 27*, 2206–2213.

Robbins, M. L., López, A. M., Weihs, K. L., & Mehl, M. R. (2014). Cancer conversations in context: Naturalistic observation of couples coping with breast cancer. *Journal of Family Psychology, 28*, 380–390.

Robbins, M. L., Mehl, M. R., Holleran, S. E., & Kasle, S. (2011). Naturalistically observed sighing and depression in rheumatoid arthritis patients: A preliminary study. *Health Psychology, 30*, 129–133.

Röcke, C., Katana, M., Fillekes, M., Martin, M., & Weibel, R. (2018). Mobility, physical activity and social interactions in the daily lives of healthy older adults: The MOASIS Project. *Innovation in Aging, 2*(Suppl 1), 274.

Rohr, M. K., John, D. T., Fung, H. H., & Lang, F. R. (2017). A three-component model of future time perspective across adulthood. *Psychology and Aging, 32*, 597–607.

Ryan, L., Hay, M., Huentelman, M. J., Duarte, A., Rundek, T., Levin, B., ... Barnes, C. (2019). Precision Aging: Applying precision medicine to the field of cognitive aging. *Frontiers in Aging Neuroscience, 11*, 128.

Salthouse, T. A. (2006). Mental exercise and mental aging: Evaluating the validity of the "use it or lose it" hypothesis. *Perspectives on Psychological Science, 1*, 68–87.

Schacter, D. L., Addis, D. R., Hassabis, D., Martin, V. C., Spreng, R. N., & Szpunar, K. K. (2012). The future of memory: Remembering, imagining, and the brain. *Neuron, 76*, 677–694.

Schmitt, M., Ringeval, F., & Schuller, B. W. (2016). At the border of acoustics and linguistics: Bag-of-audio-words for the recognition of emotions in speech. In *Interspeech* (pp. 495–499).

Schnitzspahn, K. M., Ihle, A., Henry, J. D., Rendell, P. G., & Kliegel, M. (2011). The age-prospective memory-paradox: An exploration of possible mechanisms. *International Psychogeriatrics, 23*, 583–592.

Schwarz, N. (2012). Why researchers should think "real-time": A cognitive rationale. In M. R. Mehl & T. S. Conner (Eds.), *Handbook of research methods for studying daily life* (pp. 22–42). New York, NY: The Guilford Press.

Scollon, C. N., Prieto, C. K., & Diener, E. (2009). Experience sampling: Promises and pitfalls, strength and weaknesses. In E. D. Diener (Ed.), *Assessing well-being* (pp. 157–180). Dordrecht: Springer.

Sherman, R. A., Rauthmann, J. F., Brown, N. A., Serfass, D. G., & Jones, A. B. (2015). The independent effects of personality and situations on real-time expressions of behavior and emotion. *Journal of Personality and Social Psychology, 109*, 872–888.

Song, X., & Wang, X. (2012). Mind wandering in Chinese daily lives–an experience sampling study. *PLoS One, 7*, e44423.

Stolarski, M., Fieulaine, N., & Van Beek, W. (2015). *Time perspective theory; Review, research, and application*. Cham, Switzerland: Springer.

Suddendorf, T., & Corballis, M. C. (2007). The evolution of foresight: What is mental time travel, and is it unique to humans? *Behavioral and Brain Sciences, 30*, 299–313.

Sun, J., Schwartz, H. A., Son, Y., Kern, M. L., & Vazire, S. (2019). The language of well-being: Tracking fluctuations in emotion experience through everyday speech. *Journal of Personality and Social Psychology*.

Sun, J., & Vazire, S. (2019). Do people know what they're like in the moment? *Psychological Science, 30*, 405–414.

Tackman, A. M., & Mehl, M. R. (2017). Electronically Activated Recorder (EAR). In V. Zeigler-Hill & T. K. Shackelford (Eds.), *Encyclopedia of personality and individual differences*. Cham, Switzerland: Springer.

Vazire, S. (2010). Who knows what about a person? The self–other knowledge asymmetry (SOKA) model. *Journal of Personality and Social Psychology, 98*, 281–300.

Wang, L. P., Hamaker, E., & Bergeman, C. S. (2012). Investigating inter-individual differences in short-term intra-individual variability. *Psychological Methods, 17*, 567–581.

Wank, A. A. L., Moseley, S., Polsinelli, A. J., Glisky, E. L., Mehl, M. R., & Grilli, M. D. (2019). *Out of the lab, into the real-world: Preliminary evidence that measuring autobiographical memory retrieval in a naturalistic setting replicates laboratory-based findings*. USA: Advance Online Publication.

Webster, J. D., & Ma, X. (2013). A balanced time perspective in adulthood: Well-being and developmental effects. *Canadian Journal on Aging/La Revue canadienne du vieillissement, 32*, 433–442.

Webster, J. D., Bohlmeijer, E. T., & Westerhof, G. J. (2010). Mapping the future of reminiscence: A conceptual guide for research and practice. *Research on Aging, 32*(4), 527–564.

Weidman, A. C., Sun, J., Vazire, S., Quoidbach, J., Ungar, L. H., & Dunn, E. W. (2019). (Not) hearing happiness: Predicting fluctuations in happy mood from acoustic cues using machine learning. *Emotion*.

Westerhof, G. J., Bohlmeijer, E., & Webster, J. D. (2010). Reminiscence and mental health: A review of recent progress in theory, research and interventions. *Ageing and Society, 30*, 697–721.

Wiggins, J. S. (1996). *The five-factor model of personality: Theoretical perspectives*. New York: Guilford Press.

World Health Organization. (2015). *World report on ageing and health*. Geneva, Switzerland: WHO Press.

Wrzus, C., & Mehl, M. R. (2015). Lab and/or field? Measuring personality processes and their social consequences. *European Journal of Personality, 29*, 250–271.

Wyatt, D., Choudhury, T., Bilmes, J., & Kitts, J. A. (2011). Inferring colocation and conversation networks from privacy-sensitive audio with implications for computational social science. *ACM Transactions on Intelligent Systems and Technology (TIST), 2*, 1–41.

Yordanova, K., Demiray, B., Mehl, M. R., & Martin, M. (2019). Automatic detection of everyday social behaviors and environments from verbatim transcripts of daily conversations. *IEEE International Conference on Pervasive Computing and Communications (PerCom 2019)*.

Zhang, J. W., & Howell, R. T. (2011). Do time perspectives predict unique variance in life satisfaction beyond personality traits? *Personality and Individual Differences, 50*, 1261–1266.

Zimbardo, P. G., & Boyd, J. N. (1999). Putting time in perspective: A valid, reliable individual-differences metric. *Journal of Personality and Social Psychology, 77*, 1271–1288.

Chapter 9
Exploring the Role of Mobility and Personality for Healthy Aging

Michelle Pasquale Fillekes, Camille Perchoux, Robert Weibel, and Mathias Allemand

9.1 Introduction

Promoting healthy aging has become a public health priority and key research endeavor due to rapidly aging populations around the world (WHO, 2015). Despite the decrease in physical autonomy with increasing age, factors such as independent living, maintaining an active lifestyle, and engaging in social interactions are key outcomes for both healthy aging and older adult well-being (Kestens et al., 2016; Schalock, Bonham, & Verdugo, 2008). Mobility, defined as where we move, how we move, with whom we move, why we move, and how often we move, has been found to be a key determinant of healthy aging (Cuignet et al., 2019; Hirsch, Winters, Clarke, & McKay, 2014; Musselwhite & Haddad, 2010; Schwanen & Ziegler, 2011). Mobile people are empowered to access resources which gives them a sense of autonomy (Banister & Bowling, 2004; Chung, Demiris, & Thompson,

M. P. Fillekes (✉)
Department of Geography, University of Zurich, Zurich, Switzerland

Department of Psychology, University of Zurich, Zurich, Switzerland
e-mail: michelle.fillekes@geo.uzh.ch

C. Perchoux
LISER, Luxembourg Institute of Socio-Economic Research, Esch-sur-Alzette, Luxembourg

R. Weibel
Department of Geography, University of Zurich, Zurich, Switzerland

M. Allemand
Department of Psychology, University of Zurich, Zurich, Switzerland

University Research Priority Program "Dynamics of Healthy Aging", University of Zurich, Zurich, Switzerland
e-mail: mathias.allemand@uzh.ch

P. L. Hill, M. Allemand (eds.), *Personality and Healthy Aging in Adulthood*, International Perspectives on Aging 26,
https://doi.org/10.1007/978-3-030-32053-9_9

2015; Musselwhite & Haddad, 2010). Traveling using active modes of transport (such as walking or cycling) reflects active lifestyles that have been found to correlate with physical health and well-being (Huss, Beekhuizen, Kromhout, & Vermeulen, 2014; Seresinhe, Preis, & Moat, 2015). Moreover, older adults who have access to a car or public transport, as well as good community facilities and services, have a greater number of social interactions and higher levels of well-being than their counterparts (Banister & Bowling, 2004; Gagliardi, Marcellini, Papa, Giuli, & Mollenkopf, 2010).

Besides mobility, individual differences in thinking, feeling, and behaving are important determinants of healthy aging (see the chapters in this book). This chapter focuses on individual differences in *personality traits* as dispositional tendencies and *personality states* as situational and momentary manifestations of traits in daily life (see Chap. 7 by Jackson & Beck in this book). In essence, personality traits are defined as relatively enduring and automatic patterns of behaviors, thoughts, and feelings (Baumert et al., 2017; Roberts, 2018) and describe the most basic and general dimensions upon which individuals are typically perceived to differ. Personality states, then, reflect the temporary manifestations of personality traits in response to both internal aspects, such as motives and goals, and external situations, such as stress in a given situation or interactions in a social context (Baumert et al., 2017; Hooker & McAdams, 2003). States reflect how individuals think, feel, or behave in a given situation or context. Individual differences in personality traits and states are often organized within the conceptual framework of the Big Five (John, Naumann, & Soto, 2008) or Five-Factor Model (McCrae & Costa, 2008) upon which this chapter is based as well. These models include five broad dimensions that are experienced on a spectrum: neuroticism (defined as the propensity to be anxious, worrisome, angry and depressed), extraversion (propensity to be sociable, active, assertive, and to experience positive affect), openness to experience (propensity to be original, complex, creative, and open to new ideas), agreeableness (propensity to be altruistic, trusting, modest, and warm), and conscientiousness (propensity to be self-controlled, task- and goal-directed, planful, and rule following).

The process of healthy aging is influenced by the interplay of a broad range of individual, environmental, and policy-based factors, amongst which personality and mobility can be placed (Sallis et al., 2006). Existing research has investigated the role of personality and mobility as independent determinants of individuals' health and well-being. In this chapter, however, we provide a theoretical framework for how mobility and personality together influence healthy aging. Little research exists in which relationships between mobility and personality have been investigated (Alessandretti, Lehmann, & Baronchelli, 2018; Chorley, Whitaker, & Allen, 2015; de Montjoye, Quoidbach, Robic, & Pentland, 2013). And to the best of our knowledge, no research has investigated mobility-personality interrelationships and their implications for healthy aging.

This chapter aims to combine the perspectives of mobility and personality research to help advance our understanding of healthy aging. The chapter starts by disentangling the construct of mobility by examining how it is defined, operationalized, and measured. This will serve as a foundation for the second step, in which we present a conceptual model to bridge the two domains in order to explore potential causal pathways linking mobility and personality with healthy aging. The chapter is concluded by a discussion of the model and an outlook on potential future research.

9.2 Mobility

There is an increasing number of researchers suggesting to decompose mobility into the two components *motility* (i.e., the mobility potential) and *movement* (i.e., the actual manifested mobility) and assess them separately (Fig. 9.1) (Cuignet et al., 2019; Shliselberg & Givoni, 2018; Thigpen, 2018). Motility and movement are two distinct and complementary concepts that shape an individual's mobility. In a similar fashion as how personality traits are latent and refer to an individual's personality disposition, motility is latent and refers to an individual's disposition towards mobility. Movement, on the other hand, is a context and time-dependent interpretation of mobility and thus shows parallels with the concept of personality states. Like personality traits and states, motility and movement might have differential effects on healthy aging. Both motility and movement are multidimensional concepts (dotted lines in Fig. 9.1) and each dimension can further be operationalized through multiple complementary metrics that are presented in detail in the following sections. The different aspects of both motility and movement again are expected to have potentially health-beneficial or health-damaging effects.

9.2.1 Motility: How People (Think They) Could Be Mobile

Historically rooted in biology, the concept of motility was for the first time thoroughly introduced in the social sciences by Kaufmann (2002) and defined as the capacity of a person to be mobile, or more precisely, as the way in which an individual understands what his/her possibilities are in the domain of mobility. Three core interdependent dimensions define the concept of motility: access, competence, and appropriation (Fig. 9.1) (Cuignet et al., 2019; Kaufmann, 2002; Shareck, Frohlich, & Kestens, 2014). *Access* refers to the surrounding environments shaping

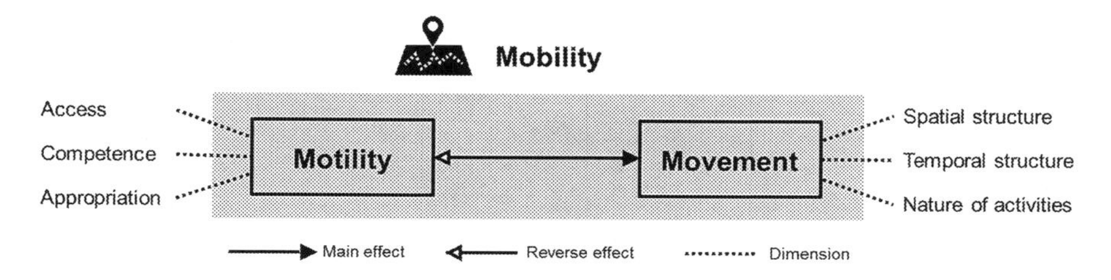

Fig. 9.1 Mobility decomposed into motility and movement with its respective dimensions

opportunities and locations with which individuals can choose to engage. *Competence* subsumes physical and cognitive abilities and skills needed to exploit the mobility options (e.g., a driver's license). Finally, *appropriation* refers to the decision-making processes and the evaluation of mobility options. Appropriation is highly influenced by a person's movement experiences, which in turn form a person's attitude towards mobility.

There is little literature in which motility was effectively operationalized with quantitative indicators (Cuignet et al., 2019; De Vos, Schwanen, Van Acker, & Witlox, 2013; Kaufmann, Dubois, & Ravalet, 2017; De Witte & Macharis, 2010). Motility is rather stable over time, as it is composed by dimensions that are typically subject to either rare changes (e.g., life changing events such as relocation to a different residential neighborhood), or slow changes (e.g., loss of physical abilities and skills, or evolution of attitudes). Consequently, motility is either assessed using single-occasion measurements as in typical cross-sectional study designs or longitudinally to capture slow developmental change, i.e., multiple times within the same individual over large temporal intervals (e.g., 4 years in Kaufmann et al., 2017). Motility is traditionally assessed using interviews (Flamm & Kaufmann, 2006) or questionnaires (Kaufmann et al., 2017). Geographic information system (GIS) based measures can be used to proxy the access dimension of motility, by qualifying and quantifying the local accessible environmental resources within an individual's residential neighborhood. Advances in mobile sensing technologies opened up new opportunities to develop more comprehensive measures of environmental exposures over an individual's entire *activity space* (e.g., Perchoux, Chaix, Cummins, & Kestens, 2013) which is defined as "the subset of all locations within which an individual has direct contact as a result of their day-to-day activities" (Golledge & Stimson, 1997, p. 279). Besides self-reports, lab-based measures can be used to infer motility-relevant competences while some of the appropriation aspects can be additionally inferred from GIS-based analysis of sensed real-life mobility data.

Because of the limited attempts to quantitatively operationalize motility, a gold standard to assess the various dimensions of motility has not yet been developed. To determine a more unified method for motility assessment, we discuss exemplary indicators for each dimension of motility (Table 9.1). Access is decomposable into indicators describing an individual's personal resources (e.g., availability of a car) and environmental resources (e.g., neighborhood walkability). Accessibility indicators describing the built environments can be grouped into the 5 'D' variables (Ewing & Cervero, 2010): density, diversity, design, destination accessibility, and distance to transit. Physical and cognitive skills constituting the competence aspect of motility essentially consist of indicators that proxy how well an individual is capable of planning and executing his/her actual movements (Giannouli, Bock, Mellone, & Zijlstra, 2016). Finally, appropriation is assessable by relying on measures that proxy mobility habits and preferences such as a participant's frequency of used transport modes and attitude towards different means of transport (Cuignet et al., 2019; De Witte, Macharis, Lannoy, & Van De Walle, 2006).

Table 9.1 Example indicators used in the literature classified into dimensions and sub-dimensions according to the motility characteristics they represent

Dimension	Indicators
Access	
Personal access	
	• Car availability / home-work distance / net income (Witte & Macharis, 2010)
	• Car / internet / computer availability (Kaufmann et al., 2017)
Environmental access	
	• Walkability in residential neighborhood, e.g., NEWS scale (Sallis, Frank, Saelens, & Kraft, 2004)
	• Characterization of residential neighborhood or activity-space-based environmental access w.r.t. the 5 'D' variables of Ewing and Cervero (2010): e.g., number of amenities (density, diversity), number of road network nodes (design of network), number of public transport stops (distance), number of buses / trains (destination accessibility) (Cuignet et al., 2019)
	• Access to highway / regional and high-speed train / airport (within 20 / 20 / 45 min, respectively) (Kaufmann et al., 2017)
Competence	
Physical skills	
	• Timed up-and-go test (Giannouli et al., 2016; Podsiadlo & Richardson, 1991)
	• Physical functioning, e.g., short-form health survey (SF-36) (Cuignet et al., 2019)
	• Bicycling skills (Thigpen, 2018)
Cognitive skills	
	• Educational level / professional status / age (Witte & Macharis, 2010)
	• Number of languages spoken (Kaufmann et al., 2017)
	• Ability to read a map and orient oneself in space (Kaufmann et al., 2017)
	• Driver's license (Cuignet et al., 2019)
	• Usage of internet (Cuignet et al., 2019)
	• Ability to consider moving more than 50 km from home (Kaufmann et al., 2017)
Appropriation	
Mobility habits / experiences	
	• Frequency of usage of different transport modes (Cuignet et al., 2019)
	• Number of regular activity places, and average number of trips per month (Cuignet et al., 2019)
Preferences / plans / attitudes	
	• Participants' attitudes regarding different aspects of different transport modes (e.g., speed, comfort, safety, commodity, ecology, etc., Cuignet et al., 2019)
	• Bicycling attitude (Thigpen, 2018)
	• Willingness to move to another region / to move abroad / to commute long distances / to commute weekly / to travel frequently on business (Kaufmann et al., 2017)

9.2.2 Movement: How People are Mobile

Besides motility, it is also important to assess the complementary concept of movement, depicting the manifestations of mobility. *Movement* is defined as the everyday spatio-temporal patterns of an individual's mobility in their environment. Movement can be analyzed by three intertwined dimensions: spatial structure, temporal structure, and the nature of activities (Fig. 9.1) (Chaix, Méline, Duncan, Jardinier, et al., 2013). Furthermore, movement determines when, where, and how people are exposed to physical (i.e., built and natural) and social environments (Chaix et al., 2012; Jankowska, Schipperijn, & Kerr, 2015; Perchoux et al., 2013; Stewart, Schipperijn, Snizek, & Duncan, 2017). In turn, movement is also influenced by the environment that surrounds individuals. The reciprocal effects between movement and daily environmental exposure, however, are not further considered in this chapter. Similarly to Pooley, Turnbull, and Adams (2005), our definition of movement refers to all travel undertaken on a temporary basis. This includes frequent and regular trips (such as the journey to work), as well as less regular but still frequent trips (to visit friends, to shop, and for other leisure activities), and trips undertaken only once or twice a year (such as visits to distant relatives).

Movement is a dynamic process that can be assessed in different ways. The classical way is to use single-occasion paper-and-pencil or online questionnaires that ask participants for their typical everyday movement at various levels of detail, such as the life-space questionnaire (Stalvey, Owsley, Sloane, & Ball, 1999) or travel diaries (Richardson, Ampt, & Meyburg, 1995). Moreover, movement can be assessed using interactive map-based questionnaires, such as the VERITAS tool used in Kestens et al. (2016). Recent studies increasingly have relied on the more objective sensor-based location sensing methods—most prominently, Global Positioning Systems (GPS)—in which participants wear sensors in custom-built devices or smartphones that continuously and unobtrusively track participants' locations (Chaix, 2018; Fillekes, Röcke, Katana, & Weibel, 2019; Hirsch et al., 2014). Self-reports typically reflect generalized information about individuals' habitual movement that often also includes additional semantic information about travel purpose or transport modes, while GPS devices collect information on movement behavior in the daily lives of participants at high spatio-temporal resolutions (Fillekes, Röcke, et al., 2019; Schipperijn et al., 2014). Often 1 week of GPS data are recorded to assess an individual's habitual movement, assuming that the majority of the movement patterns are repeated on a weekly basis (Cornwell & Cagney, 2017; Giannouli et al., 2016; Kestens et al., 2016; Schmidt, Kerr, Kestens, & Schipperijn, 2018). Some GPS studies have shown, however, that a minimum of 14 days of GPS data are needed to obtain a stable measure of an individual's activity space (Stanley, Yoo, Paul, & Bell, 2018; Zenk, Matthews, Kraft, & Jones, 2018) and people's movement habits may change to a greater extent than expected (Burkhard, Ahas, Saluveer, & Weibel, 2018).

Movement is a multi-dimensional concept, and a plethora of different movement indicators is used in the health and aging literature to assess an individual's

Table 9.2 Example movement indicators used in health/aging research classified into dimensions and sub-dimensions according to the movement characteristics they represent

Dimension	Indicators
Spatial structure	
Frequency of mobility	
	• Number of out-of-home locations (Montoliu, Blom, & Gatica-Perez, 2013)
	• Number of trips (Brusilovskiy, Klein, & Salzer, 2016; Saeb, Lattie, Schueller, Kording, & Mohr, 2016)
Extent of mobility	
	• Area of activity space (Brusilovskiy et al., 2016; Hirsch et al., 2014)
	• Average distance from home (Cornwell & Cagney, 2017; Giannouli, Bock, & Zijlstra, 2018)
	• Distance traveled (Brusilovskiy et al., 2016)
Geometry of activity space	
	• Elongation of activity space (Perchoux et al., 2014)
	• Importance of residential neighborhood with respect to entire activity space (Perchoux et al., 2014)
	• Clustering of activities (mono vs. poly-centricity) (Hasanzadeh, 2019)
Temporal structure	
Duration of mobility	
	• Travel duration (Brusilovskiy et al., 2016)
	• Time out-of-home (Brusilovskiy et al., 2016)
	• Ratio between travel time and duration spent in locations (Dijst & Vidakovic, 2000; Susilo & Dijst, 2009)
Timing / temporal rhythm	
	• Movement activities in the morning vs. the evening (Fillekes et al., in prep.)
	• Movement activities at weekends and weekdays (Kaspar, Oswald, Wahl, Voss, & Wettstein, 2015)
	• Distance from home as a function of time (Shoval et al., 2011)
Variability	
	• Day-to-day overlap in activity space (Fillekes et al., in prep.)
	• Speed variance (Saeb et al., 2016)
	• Entropy in locations (Saeb et al., 2016)
Nature of activity	
Activity diversity	
	• Number of uniquely visited locations (Brusilovskiy et al., 2016)
	• Number of different types of locations visited (Perchoux et al., 2014)
Transport mode	
	• Travel duration using active (non-motorized) vs. passive (motorized) transport modes (Fillekes et al., 2019)

movement patterns (Fillekes, Giannouli, Kim, Zijlstra, & Weibel, 2019; Perchoux et al., 2014). Commonly used movement indicators are presented in Table 9.2, classified along the three major dimensions *spatial structure, temporal structure*, and *nature of activities*, similarly as suggested in Fillekes et al. (2018, Fillekes, Giannouli, et al. 2019). Each dimension consists of further sub-dimensions that group indicators with similar characteristics.

The schematic depiction in Fig. 9.2 serves to illustrate how the movement patterns of two different individuals over 2 days can be described using indicators

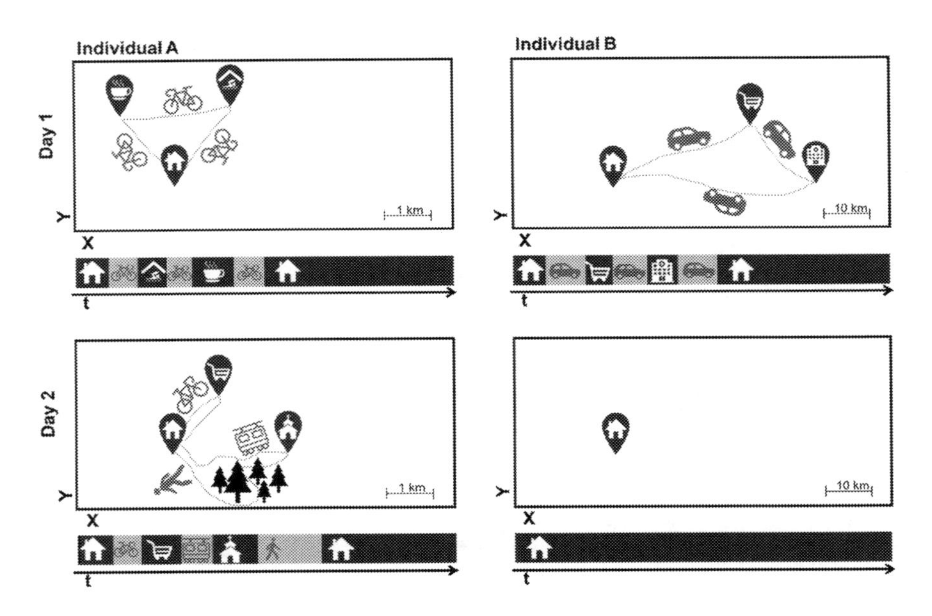

Fig. 9.2 Schematic depiction contrasting spatial (maps, X/Y) and temporal (barcode, t) view of the movement of 2 days of two individuals

reflecting differences in the movement dimensions introduced above. Individual A on average has a smaller activity space (spatial structure: extent) and more consistent activity space (temporal structure: variability), uses mostly active transport modes (nature of activity: transport mode), and visits an average of three locations per day (spatial structure: frequency). Individual B, in contrast, has more variability in the size of activity space, uses only motorized transport modes, and visits on average a lower number of locations.

9.3 A Conceptual Model Linking Mobility and Personality with Healthy Aging

An increasing number of studies have moved toward adopting more comprehensive conceptualizations of mobility and personality, respectively. Spatial science researchers suggest to integrate assessments of a person's stable motility with assessments of the dynamic construct of movement (Cuignet et al., 2019; Kaufmann, 2002; Shareck et al., 2014). Similarly, personality psychologists propose to complement the more commonly investigated relatively enduring personality traits with dynamic situation-dependent state-level fluctuations of personality (Baumert et al., 2017; Fleeson, 2004; Fleeson, Malanos, & Achille, 2002). The conceptual model (Fig. 9.3) shows the potential causal pathways linking these more comprehensive

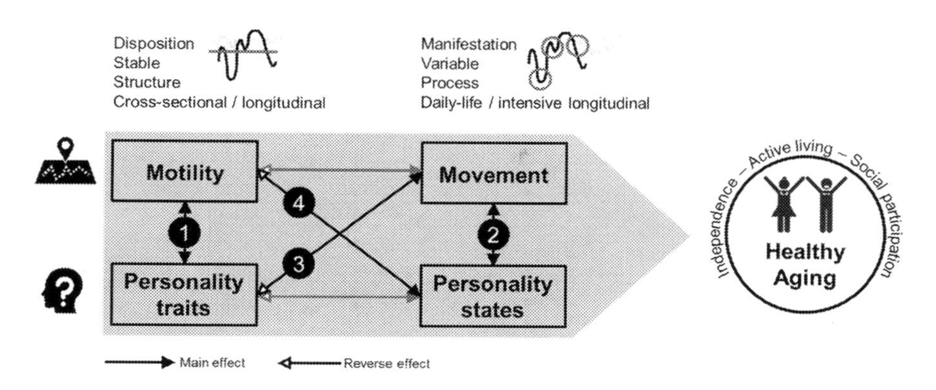

Fig. 9.3 Conceptual model illustrating the links between the components of mobility (motility and movement) and the components of personality (personality traits and states) for healthy aging

conceptualizations of mobility and personality, and healthy aging. Healthy aging is defined as the "process of developing and maintaining the functional ability that enable well-being in older age" (WHO, 2015, p. 28). The intrinsic capacity of an individual (i.e., the composite of all physical and mental capacities), the relevant environmental characteristics and the interaction between the two constitute and shape an individual's functional ability. For this chapter, we focus on the three healthy aging "outcomes" independence, active living and social participation, which are crucial domains of functional ability enabling older adults' well-being and consequently are important outcomes of healthy aging (Kestens et al., 2016; Oswald et al., 2007; Schalock et al., 2008; Webber, Porter, & Menec, 2010; WHO, 2015).

The model positions the two relatively temporally stable constructs of motility and personality traits at one end of the spectrum (left), while movement and personality states, more variable over time, are placed at the other end (right). Both motility and personality traits are reflecting the dispositional tendencies of a person with respect to the corresponding domain. With exception of rare disruptive changes (e.g., change of residence), many of the motility dimensions and personality traits are relatively enduring and show slow changes over time (Roberts, Wood, & Caspi, 2008). Cross-sectional studies typically examine whether specific motility dispositions (Cuignet et al., 2019) or certain personality traits (e.g., conscientiousness, emotional stability, or openness) (Sutin et al., 2016; Weston, Hill, & Jackson, 2015) lead to better health outcomes. Longitudinal designs are used more often in personality than in motility-related research, often to study stability or developmental change in personality traits (Roberts et al., 2008; Terracciano, Stephan, Luchetti, & Sutin, 2018). Motility and personality traits can be seen as part of an individual's functional ability (Sallis et al., 2006; WHO, 2015). Thus, they constitute important resources influencing an individual's healthy aging and well-being. Moreover, they are key structural determinants for an individual's daily life movement and personality states.

Movement and personality states describe the actual manifestations of mobility and personality, respectively. Both movement and personality states are more context-specific and time-dependent than motility and personality traits, and thus more variable over short-term intervals (Baumert et al., 2017; Fleeson, 2004). Healthy aging research that involves the daily-life assessments of movement or personality states is still relatively novel. Recent technological developments (including the miniaturization of sensors) benefit assessment strategies such as intense (i.e., high sampling intervals) longitudinal methods including sensor-based (e.g., GPS), or self-reported ambulatory assessments to describe of an individual's typical behaviors or investigate within-person processes (Allemand & Mehl, 2017; Harari, Müller, Aung, & Rentfrow, 2017; Seifert, Hofer, & Allemand, 2018). Such assessments allow to study movement and personality states of individuals which contribute to reveal the degree of people's engagement in social activities, as well as the degree to which people live actively and independently. For example, duration of active traveling has been positively associated with older adult well-being (Chung et al., 2015; Huss et al., 2014) whereas negative associations were found between long daily commutes and well-being (Stutzer & Frey, 2008). Respective findings thus far speak to the importance of each construct to healthy aging and well-being.

Overall, the conceptual model postulates that causal pathways lead mainly from the rather stable (left) via the more variable (right) constructs to healthy aging (as indicated by the "main effect" arrows in Fig. 9.3). Reverse effects (represented by respective arrows in Fig. 9.3) are less pronounced and are discussed only marginally in this chapter. Bidirectional pathways are expected for constructs that are aligned on the same level (e.g., between personality traits and motility). Regarding the link between mobility component and healthy aging, we would like to refer the reader to Cuignet et al. (2019). Associations between personality components and healthy aging are discussed in several chapters in this book. In the remaining sections, we discuss the between-domain links and how they relate to healthy aging (numbers 1 to 4 in Fig. 9.3). For each link, a subsection will be devoted to discussing existing theoretical frameworks and empirical findings, as well as hypotheses for future research.

9.3.1 Motility, Personality Traits, and Healthy Aging

An individual's motility is expected to be associated with an individual's personality traits, and both constructs individually and in combination influence the individual's healthy aging process (no. 1 in Fig. 9.3). First, regarding the access dimension of motility, personality traits may influence where people live, the characteristics of their surrounding environments and their access to environmental resources (Götz, Ebert, & Rentfrow, 2018; Jokela, Bleidorn, Lamb, Gosling, & Rentfrow, 2015). Individuals high in openness to experience and extraversion show greater tolerance for alternative lifestyles and ideas and therefore tend to reside in urban areas characterized by higher population densities and ethnic diversity

(Rentfrow, Gosling, & Potter, 2008). According to the mechanism of *self-selection* or *attraction* people choose to reside in or visit a neighborhood that fits their personality. If the fit between personality traits and motility is good, a positive effect on an individual's psychological well-being is expected (Garretsen, Stoker, Soudis, Martin, & Rentfrow, 2018; Götz et al., 2018). Similarly, in environment-health research, this mechanism is called the *selective residential bias*: individual preferences for specific environment or behavior—partly driven by personality—influence the choice of residential location (Chaix, Méline, Duncan, Merrien, et al., 2013). This choice in turn influences their health. Extraverted people, for example, might be more likely to choose to live in environments that enable social interactions, in turn promoting healthy aging. Second, regarding the competences dimension of motility, conscientiousness and openness are likely associated with a better physical and cognitive functioning (see Hill & Allemand, Chap. 1, this volume and Payne & Lohani, Chap. 11, this volume), which in turn influences the competences to be mobile, such as having a driver's license or a good physical fitness. Thereby, indirect positive impacts of personality traits are reflected in motility competences that are generally associated with an individual's sense of independence (Banister, 2018; De Vos, 2018; King et al., 2011; Rosso, Auchincloss, & Michael, 2011; Shliselberg & Givoni, 2018). Third, regarding the appropriation dimension of motility, it could be hypothesized that extraverted and open people have attitudes oriented toward active and public transport modes (Sallis et al., 2006).

Conversely, motility may have an influence on personality traits over time, as the environment they are exposed to and their mobility experiences might act on individuals' underlying personality traits (Pooley et al., 2005). In line with this idea, Zimmermann and Neyer (2013) observed that students going abroad during their studies showed increased levels of openness and extraversion and decreased levels of neuroticism—changes that are generally represented as beneficial for healthy aging (e.g., Luchetti, Terracciano, Stephan, & Sutin, 2016).

9.3.2 Movement, Personality States, and Healthy Aging

Technological advances have recently made possible the study of personality states in daily life with different methods that can be summarized under the term ecological momentary assessment (see Demiray et al., Chap. 8, this volume; Jackson & Beck, Chap. 7, this volume). This includes using mobile phone apps for self-reporting or observational methods including audio and video recording, or analyzing online social network data or mobile phone call data (Allemand & Mehl, 2017; Toole, Herrera-Yaque, Schneider, & Gonzalez, 2015). Combining these methods with location-sensing technologies holds the potential for understanding how personality states and movement patterns influence each other, and further impact health (see no. 2 in Fig. 9.3).

To our knowledge, the sole study investigating the interface between movement and social behaviors is the one of Alessandretti et al. (2018). They found that the size and stability of activity space over time are related to the size and stability of an individual's social network. They suggest that these manifestations of movement are explained by similar dispositions in the personality traits of extraversion, openness, and neuroticism. Another potential explanation is that individuals who have a higher number of contacts have to travel greater distances in order to maintain their social contacts. Older adults who manifest many extraverted and open states, and who engage with a large activity space are likely to be more independent and engaged in social participation (c.f. Viljanen, Mikkola, Rantakokko, Portegijs, & Rantanen, 2015). Conversely, traveling using active and public means of transport might stimulate open and extraverted personality states that have reinforcing beneficial effects on healthy aging.

9.3.3 Personality Traits, Movement, and Healthy Aging

Research using personality traits to explain individuals' differing movement patterns is more widespread and comes to the following conclusions (see no. 3 in Fig. 9.3): More open, extraverted and emotionally stable people have higher numbers of visited locations, larger distances travelled, and show less routine behavior than individuals with lower propensities in these traits (Alessandretti et al., 2018; Chorley et al., 2015; de Montjoye et al., 2013). Personality traits indirectly influence healthy aging, by triggering movement patterns positively associated with social participation, active living and well-being. The same traits that are known to shape an individual's engagement with physical activity and that are generally known to be associated with healthy behaviors—i.e., high levels of extraversion, openness, conscientiousness and a low level of neuroticism (e.g., Bogg & Roberts, 2004; Sutin et al., 2016)—are likely to foster the use of more health-beneficial active transport modes (e.g., walking and bicycle) and less motorized ones, known to negatively affect momentary well-being (Seresinhe, Preis, MacKerron, & Moat, 2019). However, using personality traits to explain patterns of movement is still a recent endeavor. Future research should link personality traits to specific types of movement patterns, which in turn may promote or impair the process of healthy aging.

Besides the study of the indirect effects of personality traits on healthy aging via movement, the study of personality traits as potential moderators between movement and healthy aging predictors offers an exciting field of research. One could, for example, evaluate whether open or extraverted individuals experience greater well-being after moving through a range of public locations. If this was the case, these traits may explain to what degree an individual takes advantage of the possibility to engage with the environment offered through the visited locations, and thus has beneficial effects on their well-being and healthy aging.

9.3.4 Motility, Personality States, and Healthy Aging

Motility offers interesting insights into how an individual's mobility context influences their manifested personality states (see no. 4 in Fig. 9.3). Besides a direct impact of motility on healthy aging, such as living in a residential neighborhood with good accessibility to transport and various facilities having positive impacts on subjective well-being (Liu, Dijst, & Geertman, 2017), motility might have indirect impacts on healthy aging by fostering certain health-beneficial personality states. For example, older adults living in neighborhoods characterized by good walkability and opportunities for social engagements may show more extraverted and open personality states in their daily lives, and in turn potentially higher levels of well-being. In the opposite, living in a more deprived and unsafe neighborhood may lead to more neurotic states such as anxiety and might consequently decrease well-being. Though studying motility and personality states in isolation already may uncover some healthy aging relevant relationships, integrating these constructs with their respective counterparts (i.e., movement and personality traits) could offer a higher potential to understand the mechanisms at play in shaping healthy aging.

9.4 Additional Comments on the Conceptual Model

This section addresses how to combine comprehensive views of both mobility and personality in healthy aging research. Furthermore, we attempt to show how this combination can help to identify populations at risk for adverse outcomes based on their mobility and personality profiles, and to tailor interventions that simultaneously target the more structural and variable constructs to have maximal health-beneficial effects for aging.

Including mobility into models of personality and healthy aging might explain why older adults with similar personality traits might show different health-beneficial personality states. Some people might be trapped in a deprived neighborhood and are constrained by their accessibility or mobility competences, and consequently show little movement, which results in less healthy personality states. By contrast, people who can move from their residence to another neighborhood that better fits their personality might be in better health. Moreover, motility might help to explain why an extraverted individual might show introverted states when traveling by train; namely, the person might simply be not accustomed to the use of public means of transport. Investigating whether specific motility-movement configurations are supportive for certain personalities or foster health-beneficial personality states could help promote healthy aging. In this logic, the suggestion to vary the daily travel itineraries would be appropriate for a person high in openness and extraversion but inappropriate for a person low on the respective traits. As older adults generally become less mobile in terms of both motility and movement, their residential neighborhood derives a higher symbolic and contextual importance

(Vallée, Cadot, Roustit, Parizot, & Chauvin, 2011). Hence, achieving a good fit between personality and residential neighborhood is even more important. Research along these lines might help uncover how to design environments supportive of healthy aging taking into account individuals' preferences and needs derived from their personality profiles (Götz et al., 2018; Rosso et al., 2011).

Including personality into mobility-healthy aging research might contribute to explaining why older adults with similar mobility potentials manifest different movement behavior. People high in openness and extraversion, for example, may be more likely to travel beyond their residential neighborhoods because they might feel more at ease with using public means of transport. The resulting travel experience might expose older adults to different environments and allows them to access the goods and resources outside their neighborhood. These are behaviors that aid the promotion of healthy aging by fostering independence, active lifestyles and social participation. Tailored interventions targeting older adults without a driver's license, who also happen to be low in openness and extraversion, might consist of organized age-specific transport facilities to create a more familiar setting, as opposed to public transport (Viljanen et al., 2015). Moreover, individual differences in personality traits cause different desires and needs for movement (Chorley et al., 2015; Götz et al., 2018; Jokela et al., 2015; de Montjoye et al., 2013).

Exploring the different components of both mobility and personality is also relevant with respect to different temporal scales of analysis: Motility and personality traits might be more relevant for predicting global quality of life and well-being, as they are closely linked with an individual's intrinsic capacities. Whereas movement and personality states might lend themselves better to the study of the more short-term subjective well-being and its fluctuations. Thus, future research should investigate how motility and personality traits help explain an individual's momentary well-being. For example, individuals might feel uncomfortable traveling with public means of transport if their mobility and personality resources do not meet the activity's requirements (e.g., being familiar with using public transport modes, and having open, and extraverted personality traits). Knowing that an individual is low on the openness spectrum and feels uncomfortable in new situations, strategies could be developed to help familiarize the individual with the public transport system through an online tutorial to empower them to navigate in public transport with greater ease (i.e., improving the individual's motility). A better understanding of the causal pathways linking all these constructs to healthy aging would result in public health interventions that are more efficient by focusing on intervening on the constructs that have optimal direct and indirect positive effects on healthy aging outcomes, or by targeting a specific sub-population for which positive effects of an intervention are to be expected.

9.5 Challenges and Future Directions

Assessing the multidimensional nature of motility, movement, personality traits and states is important because the individual dimensions of each construct are likely to have complementary and differential effects on healthy aging. When it comes to assessing movement, however, more research is required to identify the fundamental underlying dimensions of movement (Fillekes, Giannouli, et al., 2019; Perchoux et al., 2014). There are also technical issues still remaining. For instance, despite a large number of recent studies that are based on GPS assessments, limited battery life, as well as dealing with erroneous or missing GPS signal in and around buildings, still pose problems (Kerr, Duncan, & Schipperjin, 2011). Lastly, also more consensus on how to quantitatively operationalize motility will help the field to advance (Cuignet et al., 2019; Flamm & Kaufmann, 2006; Kaufmann et al., 2017).

Regarding the directionalities of the causal pathways between the constructs, many of the examples are leading from the dispositions via the manifestations to the healthy aging outcomes. However, the potential for reverse causality should also be further investigated using prospective longitudinal studies or quasi-experimental designs. Movement and personality states that foster independent and active living, and social participation, certainly contribute to maintaining a positive well-being as one enters older adulthood. Conversely, health status is also expected to condition what types of movement and personality states are manifested. Furthermore, the proposed conceptual model focused on mobility and personality. However, other factors such as socioeconomic conditions, political and cultural circumstances may have an impact on healthy aging as well and may interact with mobility and personality (Pooley et al., 2005; Sallis et al., 2006).

A next step to an understanding of the mechanisms determining how mobility and personality impact on healthy aging would entail the formulation and empirical testing of hypotheses on the potential mediating and moderating effects of each dimension of the mobility and personality components on each other as well as on healthy aging outcomes. To answer questions exploring the interplay of motility, movement, personality traits, and states, new types of study designs are needed that combine comprehensive measurements of the stable constructs with intensive bursts of daily-life assessments of the more fluctuating constructs. Subsequently, statistical methods such as longitudinal structural equation modeling (Little, 2013) and multilevel modeling (Bolger & Laurenceau, 2013) might be helpful to test the validity of the conceptual model. Motility and personality traits are important resources determining what competences and environments older adults have at their disposal to engage in activities representing active and independent living, and engagement in social interaction. Healthy aging could for example be fostered by enhancing an individual's motility and therefore influencing an individual's healthy aging process via multiple pathways. Example health interventions could be targeted at improving access to transport facilities or to change people's attitudes towards active transport modes by promoting its ecological and healthy aspects. Such interventions directly positively influence physical health, but also indirectly positively act on healthy aging by facilitating more out-of-home movements, which in turn stimulate open and extraverted personality states. Future research looking into the relations between mobility, personality, and healthy aging will help to further develop theory and reveal how healthy aging can be promoted based on individuals' mobility and personality.

References

Alessandretti, L., Lehmann, S., & Baronchelli, A. (2018). Individual mobility and social behaviour: Two sides of the same coin. *ArXiv Preprint ArXiv, 1801*(03962), 1–31. Retrieved from http://arxiv.org/abs/1801.03962

Allemand, M., & Mehl, M. R. (2017). Personality assessment in daily life: A roadmap for future personality development research. In *Personality development across the lifespan* (pp. 437–454). San Diego, CA: Elsevier. https://doi.org/10.1016/B978-0-12-804674-6.00027-2

Banister, D. (2018). *Inequality in transport*. Oxford, UK: Alexandrine Press.

Banister, D., & Bowling, A. (2004). Quality of life for the elderly: The transport dimension. *Transport Policy, 11*(2), 105–115. https://doi.org/10.1016/S0967-070X(03)00052-0

Baumert, A., Schmitt, M., Perugini, M., Johnson, W., Blum, G., Borkenau, P., … Wrzus, C. (2017). Integrating personality structure, personality process, and personality development. *European Journal of Personality, 31*(5), 503–528. https://doi.org/10.1002/per.2115

Bogg, T., & Roberts, B. W. (2004). Conscientiousness and health-related behaviors: A meta-analysis of the leading behavioral contributors to mortality. *Psychological Bulletin, 130*(6), 887–919. https://doi.org/10.1037/0033-2909.130.6.887

Bohte, W., & Maat, K. (2009). Deriving and validating trip purposes and travel modes for multi-day GPS-based travel surveys: A large-scale application in the Netherlands. *Transportation Research Part C: Emerging Technologies, 17*(3), 285–297. https://doi.org/10.1016/j.trc.2008.11.004

Bolger, N., & Laurenceau, J.-P. (2013). *Intensive longitudinal methods: An introduction to diary and experience sampling research*. New York, NY: Guilford Press.

Brusilovskiy, E., Klein, L. A., & Salzer, M. S. (2016). Using global positioning systems to study health-related mobility and participation. *Social Science and Medicine, 161*, 134–142. https://doi.org/10.1016/j.socscimed.2016.06.001

Burkhard, O., Ahas, R., Saluveer, E., & Weibel, R. (2018). Extracting regular mobility patterns from sparse CDR data without a priori assumptions. *Journal of Location Based Services, 11*(2), 78–97. https://doi.org/10.1080/17489725.2017.1333638

Chaix, B. (2018). Mobile sensing in environmental health and neighborhood research. *Annual Review of Public Health, 39*, 367–384. https://doi.org/10.1146/annurev-publhealth-040617-013731

Chaix, B., Kestens, Y., Perchoux, C., Karusisi, N., Merlo, J., & Labadi, K. (2012). An interactive mapping tool to assess individual mobility patterns in neighborhood studies. *American Journal of Preventive Medicine, 43*(4), 440–450. https://doi.org/10.1016/j.amepre.2012.06.026

Chaix, B., Méline, J., Duncan, S., Jardinier, L., Perchoux, C., Vallée, J., … Kestens, Y. (2013). Neighborhood environments, mobility, and health: Towards a new generation of studies in environmental health research. *Revue d'Épidémiologie et de Santé Publique, 61*, 139–145. https://doi.org/10.1016/j.respe.2013.05.017

Chaix, B., Méline, J., Duncan, S., Merrien, C., Karusisi, N., Perchoux, C., … Kestens, Y. (2013). GPS tracking in neighborhood and health studies: A step forward for environmental exposure assessment, a step backward for causal inference? *Health & Place, 21*, 46–51. https://doi.org/10.1016/j.healthplace.2013.01.003

Chorley, M. J., Whitaker, R. M., & Allen, S. M. (2015). Personality and location-based social networks. *Computers in Human Behavior, 46*, 45–56. https://doi.org/10.1016/j.chb.2014.12.038

Chung, J., Demiris, G., & Thompson, H. J. (2015). Instruments to assess mobility limitation in community-dwelling older adults: A systematic review. *Journal of Aging and Physical Activity, 23*(2), 298–313. https://doi.org/10.1123/japa.2013-0181

Cornwell, E. Y., & Cagney, K. A. (2017). Aging in activity space: Results from smartphone-based GPS-tracking of urban seniors. *Journals of Gerontology: Social Sciences, 72*(5), 864–875. https://doi.org/10.1093/geronb/gbx063

Cuignet, T., Perchoux, C., Caruso, G., Klein, O., Klein, S., Chaix, B., … Gerber, P. (2019). Mobility among older adults: Deconstructing motility and movement effects on wellbeing. *Urban Studies*. https://doi.org/10.1177/0042098019852033

de Montjoye, Y.-A., Quoidbach, J., Robic, F., & Pentland, A. S. (2013). Predicting personality using novel mobile phone-based metrics. In *International conference on social computing, behavioral-cultural modeling, and predictionrediction* (pp. 48–55). Berlin/Heidelberg, Germany: Springer. https://doi.org/10.1007/978-3-642-37210-0_6

De Vos, J. (2018). Towards happy and healthy travellers: A research agenda. *Journal of Transport & Health, 11*, 80–85. https://doi.org/10.1016/J.JTH.2018.10.009

De Vos, J., Schwanen, T., Van Acker, V., & Witlox, F. (2013). Travel and subjective well-being: A focus on findings, methods and future research needs. *Transport Reviews, 33*(4), 421–442. https://doi.org/10.1080/01441647.2013.815665

De Witte, A., & Macharis, C. (2010). Commuting to Brussels: How attractive is "free" public transport? *Brussels Studies, 32*(37), 0–16. https://doi.org/10.4000/brussels.755

De Witte, A., Macharis, C., Lannoy, P., & Van De Walle, S. (2006). The impact of "'free'" public transport: The case of Brussels. *Transportation Research Part A: Policy and Practice, 40*, 671–689. https://doi.org/10.1016/j.tra.2005.12.008

Dijst, M., & Vidakovic, V. (2000). Travel time ratio: The key factor of spatial reach. *Transportation, 27*, 179–199. Retrieved from https://doi.org/10.1023/a:1005293330869.pdf

Ewing, R., & Cervero, R. (2010). Travel and the built environment. *Journal of the American Planning Association, 76*(3), 265–294. https://doi.org/10.1080/01944361003766766

Fillekes, M. P., Giannouli, E., Kim, E.-K., Zijlstra, W., & Weibel, R. (2019). Towards a comprehensive set of GPS-based indicators reflecting the multidimensional nature of daily mobility for applications in health and aging research. *International Journal of Health Geographics, 18*(17). https://doi.org/10.1186/s12942-019-0181-0

Fillekes, M. P., Giannouli, E., Zijlstra, W., & Weibel, R. (2018). Towards a framework for assessing daily mobility using GPS data. *GI_Forum, 1*, 177–183. https://doi.org/10.1553/giscience2018_01_s177

Fillekes, M. P., Röcke, C., Katana, M., & Weibel, R. (2019). Self-reported versus GPS-derived indicators of daily mobility in a sample of healthy older adults. *Social Science & Medicine, 220*, 193–202. https://doi.org/10.1016/j.socscimed.2018.11.010

Flamm, M., & Kaufmann, V. (2006). Operationalising the concept of motility: A qualitative study. *Mobilities, 1*(2), 167–189. https://doi.org/10.1080/17450100600726563

Fleeson, W. (2004). Moving personality beyond the person-situation debate. *Current Directions in Psychological Science, 13*(2), 83–87.

Fleeson, W., Malanos, A. B., & Achille, N. M. (2002). An intraindividual process approach to the relationship between extraversion and positive affect: Is acting extraverted as "good" as being extraverted? *Journal of Personality and Social Psychology, 83*(6), 1409–1422. https://doi.org/10.1037//0022-3514.83.6.1409

Gagliardi, C., Marcellini, F., Papa, R., Giuli, C., & Mollenkopf, H. (2010). Associations of personal and mobility resources with subjective well-being among older adults in Italy and Germany. *Archives of Gerontology and Geriatrics, 50*(1), 42–47. https://doi.org/10.1016/j.archger.2009.01.007

Garretsen, H., Stoker, J. I., Soudis, D., Martin, R., & Rentfrow, J. (2018). The relevance of personality traits for urban economic growth: Making space for psychological factors. *Journal of Economic Geography*, 1–25. https://doi.org/10.1093/jeg/lby025

Giannouli, E., Bock, O., Mellone, S., & Zijlstra, W. (2016). Mobility in old age: Capacity is not performance. *BioMed Research International, February*, 1–8. https://doi.org/10.1155/2016/3261567

Giannouli, E., Bock, O., & Zijlstra, W. (2018). Cognitive functioning is more closely related to real-life mobility than to laboratory-based mobility parameters. *European Journal of Ageing*, 1–9. https://doi.org/10.1007/s10433-017-0434-3

Golledge, R. G., & Stimson, R. J. (1997). *Spatial behavior: A geographic perspective*. New York, NY: Guilford Press.

Götz, F. M., Ebert, T., & Rentfrow, P. J. (2018). Regional cultures and the psychological geography of Switzerland: Person–environment–fit in personality predicts subjective wellbeing. *Frontiers in Psychology, 9*, 517. https://doi.org/10.3389/fpsyg.2018.00517

Harari, G. M., Müller, S. R., Aung, M. S., & Rentfrow, P. J. (2017). Smartphone sensing methods for studying behavior in everyday life. *Current Opinion in Behavioral Sciences, 18*, 83–90. https://doi.org/10.1016/j.cobeha.2017.07.018

Hasanzadeh, K. (2019). Exploring centricity of activity spaces: From measurement to the identification of personal and environmental factors. *Travel Behaviour and Society, 14*, 57–65. https://doi.org/10.1016/j.tbs.2018.10.001

Hooker, K., & McAdams, D. P. (2003). Personality reconsidered: A new agenda for aging research. *The Journals of Gerontology Series B: Psychological Sciences and Social Sciences, 58*(6), 296–304

Hirsch, J. A., Winters, M., Clarke, P., & McKay, H. (2014). Generating GPS activity spaces that shed light upon the mobility habits of older adults: A descriptive analysis. *International Journal of Health Geographics, 13*(1), 51. https://doi.org/10.1186/1476-072X-13-51

Huss, A., Beekhuizen, J., Kromhout, H., & Vermeulen, R. (2014). Using GPS-derived speed patterns for recognition of transport modes in adults. *International Journal of Health Geographics, 13*(1), 40. https://doi.org/10.1186/1476-072X-13-40

Jankowska, M., Schipperijn, J., & Kerr, J. (2015). A framework for using GPS data in physical activity and sedentary behavior studies. *Exercise and Sport Sciences Reviews, 43*(1), 48–56.

John, O. P., Naumann, L. P., & Soto, C. J. (2008). Paradigm shift to the integrative Big Five Trait taxonomy. In O. P. John, R. Robins, & L. Pervin (Eds.), *Handbook of personality: Theory and research 3.2* (3rd ed., pp. 114–158). New York, NY: Guilford Publications. https://doi.org/10.1016/S0191-8869(97)81000-8

Jokela, M., Bleidorn, W., Lamb, M. E., Gosling, S. D., & Rentfrow, P. J. (2015). Geographically varying associations between personality and life satisfaction in the London metropolitan area. *Proceedings of the National Academy of Sciences, 112*(3), 725–730. https://doi.org/10.1073/pnas.1415800112

Kaspar, R., Oswald, F., Wahl, H.-W., Voss, E., & Wettstein, M. (2015). Daily mood and out-of-home mobility in older adults: Does cognitive impairment matter? *Journal of Applied Gerontology, 34*(1), 26–47. https://doi.org/10.1177/0733464812466290

Kaufmann, V. (2002). *Re-thinking mobility: Contemporary sociology (Transport and Society)* (1st ed.). New York, NY: Routledge.

Kaufmann, V., Dubois, Y., & Ravalet, E. (2017). Measuring and typifying mobility using motility. *Applied Mobilities, 0127*, 1–16. https://doi.org/10.1080/23800127.2017.1364540

Kerr, J., Duncan, S., & Schipperjin, J. (2011). Using global positioning systems in health research: A practical approach to data collection and processing. *American Journal of Preventive Medicine, 41*(5), 532–540. https://doi.org/10.1016/j.amepre.2011.07.017

Kestens, Y., Chaix, B., Gerber, P., Desprès, M., Gauvin, L., Klein, O., … Wasfi, R. (2016). Understanding the role of contrasting urban contexts in healthy aging: An international cohort study using wearable sensor devices (the CURHA study protocol). *BMC Geriatrics, 16*(1), 1–12. https://doi.org/10.1186/s12877-016-0273-7

Kestens, Y., Wasfi, R., Naud, A., & Chaix, B. (2017). "Contextualizing context": Reconciling environmental exposures, social networks, and location preferences in health research. *Curr Envir Health Rpt, 4*, 51–60. https://doi.org/10.1007/s40572-017-0121-8

King, A. C., Sallis, J. F., Frank, L. D., Saelens, B. E., Cain, K., Conway, T. L., … Kerr, J. (2011). Aging in neighborhoods differing in walkability and income: Associations with physical activity and obesity in older adults. *Social Science & Medicine, 73*, 1525–1533. https://doi.org/10.1016/j.socscimed.2011.08.032

Little, T. D. (2013). *Longitudinal structural equation modeling*. New York, NY: The Guilford Press.

Liu, Y., Dijst, M., & Geertman, S. (2017). The subjective well-being of older adults in Shanghai: The role of residential environment and individual resources. *Urban Studies, 54*(7), 1692–1714. https://doi.org/10.1177/0042098016630512

Luchetti, M., Terracciano, A., Stephan, Y., & Sutin, A. R. (2016). Personality and cognitive decline in older adults: Data from a longitudinal sample and meta-analysis. *The Journals of Gerontology Series B: Psychological Sciences and Social Sciences, 71*(4), 591–601. https://doi.org/10.1093/geronb/gbu184

McCrae, R. R., & Costa, P. T. (2008). Empirical and theoretical status of the five-factor model of personality traits. *The SAGE Handbook of Personality Theory and Assessment, 1*(January), 273–294. https://doi.org/10.4135/9781849200462.n13

Montoliu, R., Blom, J., & Gatica-Perez, D. (2013). Discovering places of interest in everyday life from smartphone data. *Multimedia Tools and Applications, 62*(1), 179–207. https://doi.org/10.1007/s11042-011-0982-z

Musselwhite, C., & Haddad, H. (2010). Mobility, accessibility and quality of later life. *Quality in Ageing and Older Adults, 11*(1), 25–37. https://doi.org/10.5042/qiaoa.2010.0153

Oswald, F., Wahl, H.-W., Schilling, O., Nygren, C., Fänge, A., Sixsmith, A., … Iwarsson, S. (2007). Relationship between housing and healthy aging in very old age. *The Gerontologist, 47*(1), 96–107.

Perchoux, C., Brondeel, R., Wasfi, R., Klein, O., Caruso, G., Vallée, J., … Gerber, P. (2019). Walking, trip purpose, and exposure to multiple environments: A case study of older adults in Luxembourg. *Journal of Transport & Health, 13*, 170–184. https://doi.org/10.1016/j.jth.2019.04.002

Perchoux, C., Chaix, B., Cummins, S., & Kestens, Y. (2013). Conceptualization and measurement of environmental exposure in epidemiology: Accounting for activity space related to daily mobility. *Health & Place, 21*, 86–93. https://doi.org/10.1016/j.healthplace.2013.01.005

Perchoux, C., Kestens, Y., Thomas, F., Van Hulst, A., Thierry, B., & Chaix, B. (2014). Assessing patterns of spatial behavior in health studies: Their socio-demographic determinants and associations with transportation modes (the RECORD Cohort Study). *Social Science and Medicine, 119*, 64–73. https://doi.org/10.1016/j.socscimed.2014.07.026

Podsiadlo, D., & Richardson, S. (1991). The Timed "Up and Go": A test of basic functional mobility for frail elderly persons. *Journal of the American Geriatrics Society*, 39(2), 142–148. https://doi.org/10.1111/j.1532-5415.1991.tb01616.x

Pooley, C., Turnbull, J., & Adams, M. (2005). In B. Graham (Ed.), *A mobile century?: Changes in everyday mobility in Britain in the twentieth century*. London, UK: Routledge.

Rentfrow, P. J., Gosling, S. D., & Potter, J. (2008). A theory of the emergence, persistence, and expression of geographic variation in psychological characteristics. *Perspectives on Psychological Science, 3*(5), 339–369. https://doi.org/10.1111/j.1745-6924.2008.00084.x

Richardson, A. J., Ampt, E. S., & Meyburg, A. H. (1995). *Survey methods in transport planning*. Melbourne, Australia: Eucalyptus Press.

Roberts, B. W. (2018). A revised sociogenomic model of personality traits. *Journal of Personality, 86*(1), 23–35. https://doi.org/10.1111/jopy.12323

Roberts, B. W., Wood, D., & Caspi, A. (2008). The development of personality traits in adulthood. In P. Oliver, R. W. R. John, & L. A. Pervin (Eds.), *Handbook of personality: Theory and research* (3rd ed., pp. 375–398). New York, NY: The Guilford Press.

Rosso, A. L., Auchincloss, A. H., & Michael, Y. L. (2011). The urban built environment and mobility in older adults: A comprehensive review. *Journal of Aging Research, 2011*(Article ID 816106), 1–10. https://doi.org/10.4061/2011/816106
Saeb, S., Lattie, E. G., Schueller, S. M., Kording, K. P., & Mohr, D. C. (2016). The relationship between mobile phone location sensor data and depressive symptom severity. *PeerJ, 4*, e2537. https://doi.org/10.7717/peerj.2537
Sallis, J. F., Cervero, R. B., Ascher, W., Henderson, K. A., Kraft, M. K., & Kerr, J. (2006). An ecological approach to creating active living communities. *Annual Review of Public Health, 27*, 297–322. https://doi.org/10.1146/annurev.publhealth.27.021405.102100
Sallis, J. F., Frank, L. D., Saelens, B. E., & Kraft, M. K. (2004). Active transportation and physical activity: Opportunities for collaboration on transportation and public health research. *Transportation Research Part A: Policy and Practice, 38*(4), 249–268. https://doi.org/10.1016/j.tra.2003.11.003
Schalock, R. L., Bonham, G. S., & Verdugo, M. A. (2008). The conceptualization and measurement of quality of life: Implications for program planning and evaluation in the field of intellectual disabilities. *Evaluation and Program Planning, 31*(2), 181–190. https://doi.org/10.1016/j.evalprogplan.2008.02.001
Schipperijn, J., Kerr, J., Duncan, S., Madsen, T., Klinker, C. D., & Troelsen, J. (2014). Dynamic accuracy of GPS receivers for use in health research: A novel method to assess GPS accuracy in real-world settings. *Frontiers in Public Health, 2*. https://doi.org/10.3389/fpubh.2014.00021
Schmidt, T., Kerr, J., Kestens, Y., & Schipperijn, J. (2018). Challenges in using wearable GPS devices in low-income older adults: Can map-based interviews help with assessments of mobility? *Translational Behavioral Medicine*. https://doi.org/10.1093/tbm/iby009
Schwanen, T., & Ziegler, F. (2011). Wellbeing, independence and mobility: An introduction. *Ageing and Society, 2011*, 719–733. https://doi.org/10.1017/S0144686X10001467
Seifert, A., Hofer, M., & Allemand, M. (2018). Mobile data collection: Smart, but not (yet) smart enough. *Frontiers in Neuroscience, 12*, 971. https://doi.org/10.3389/fnins.2018.00971
Seresinhe, C. I., Preis, T., MacKerron, G., & Moat, H. S. (2019). Happiness is greater in more scenic locations. *Scientific Reports, 9*(1), 4498. https://doi.org/10.1038/s41598-019-40854-6
Seresinhe, C. I., Preis, T., & Moat, H. S. (2015). Quantifying the impact of scenic environments on health. *Scientific Reports, 5*(Article number 16899), 1–9. https://doi.org/10.1038/srep16899
Shareck, M., Frohlich, K. L., & Kestens, Y. (2014). Considering daily mobility for a more comprehensive understanding of contextual effects on social inequalities in health: A conceptual proposal. *Health & Place, 29*, 154–160. https://doi.org/10.1016/j.healthplace.2014.07.007
Shliselberg, R., & Givoni, M. (2018). Motility as a policy objective. *Transport Reviews, 38*(3), 279–297. https://doi.org/10.1080/01441647.2017.1355855
Shoval, N., Wahl, H.-W., Auslander, G., Isaacson, M., Oswald, F., Edry, T., … Heinik, J. (2011). Use of the global positioning system to measure the out-of-home mobility of older adults with differing cognitive functioning. *Ageing and Society, 31*(05), 849–869. https://doi.org/10.1017/S0144686X10001455
Stalvey, B. T., Owsley, C., Sloane, M. E., & Ball, K. (1999). The life space questionnaire: A measure of the extent of mobility of older adults. *Journal of Applied Gerontology, 18*(4), 460–478. https://doi.org/10.1177/073346489901800404
Stanley, K., Yoo, E.-H., Paul, T., & Bell, S. (2018). How many days are enough?: Capturing routine human mobility. *International Journal of Geographical Information Science, 32*(6). https://doi.org/10.1080/13658816.2018.1434888
Stewart, T., Schipperijn, J., Snizek, B., & Duncan, S. (2017). Adolescent school travel: Is online mapping a practical alternative to GPS-assessed travel routes? *Journal of Transport and Health, 5*, 113–122. https://doi.org/10.1016/j.jth.2016.10.001
Stutzer, A., & Frey, B. S. (2008). Stress that doesn't pay: The commuting paradox ∗. *The Scandinavian Journal of Economics, 110*(2), 339–366. https://doi.org/10.1111/j.1467-9442.2008.00542.x

Susilo, Y. O., & Dijst, M. (2009). How far is too far? travel time ratios for activity participation in the Netherlands. *Transportation Research Record: Journal of the Transportation Research, 2134*, 89–98. https://doi.org/10.3141/2134-11

Sutin, A. R., Stephan, Y., Luchetti, M., Artese, A., Oshio, A., & Terracciano, A. (2016). The five-factor model of personality and physical inactivity: A meta-analysis of 16 samples. *Journal of Research in Personality, 63*, 22–28. https://doi.org/10.1016/J.JRP.2016.05.001

Terracciano, A., Stephan, Y., Luchetti, M., & Sutin, A. R. (2018). Cognitive impairment, dementia, and personality stability among older adults. *Assessment, 25*(3), 336–347. https://doi.org/10.1177/1073191117691844

Thigpen, C. (2018). Do bicycling experiences and exposure influence bicycling skills and attitudes? Evidence from a bicycle-friendly university. *Transportation Research Part A: Policy and Practice, 110*, 189–190. https://doi.org/10.1016/j.tra.2018.05.017

Toole, J. L., Herrera-Yaque, C., Schneider, C. M., & Gonzalez, M. C. (2015). Coupling human mobility and social ties. *Journal of the Royal Society Interface, 12*(105), 20141128. https://doi.org/10.1098/rsif.2014.1128

Vallée, J., Cadot, E., Roustit, C., Parizot, I., & Chauvin, P. (2011). The role of daily mobility in mental health inequalities: The interactive influence of activity space and neighbourhood of residence on depression. *Social Science & Medicine, 73*(8), 1133–1144. https://doi.org/10.1016/j.socscimed.2011.08.009

Viljanen, A., Mikkola, T. M., Rantakokko, M., Portegijs, E., & Rantanen, T. (2015). The association between transportation and life-space mobility in community-dwelling older people with or without walking difficulties. *Journal of Aging and Health, 35*, 1–17. https://doi.org/10.1177/0898264315618919

Webber, S. C., Porter, M. M., & Menec, V. H. (2010). Mobility in older adults: A comprehensive framework. *Gerontologist, 50*(4), 443–450. https://doi.org/10.1093/geront/gnq013

Weston, S. J., Hill, P. L., & Jackson, J. J. (2015). Personality traits predict the onset of disease. *Social Psychological and Personality Science, 6*(3), 309–317. https://doi.org/10.1177/1948550614553248

WHO. (2015). World report on ageing and health. *World Health Organization*. Retrieved from http://www.who.int/ageing/events/world-report-2015-launch/en/

Zenk, S. N., Matthews, S. A., Kraft, A. N., & Jones, K. K. (2018). How many days of global positioning system (GPS) monitoring do you need to measure activity space environments in health research? *Health and Place, 51*(October 2017), 52–60. https://doi.org/10.1016/j.healthplace.2018.02.004

Zimmermann, J., & Neyer, F. J. (2013). Do we become a different person when hitting the road? Personality development of sojourners. *Journal of Personality and Social Psychology, 105*(3), 515–530. https://doi.org/10.1037/a0033019

Chapter 10
Promoting Cognitive, Physical, and Social Activities for Healthy Aging by Targeting Personality

Damaris Aschwanden and Mathias Allemand

10.1 Introduction

Organizations such as the World Health Organization (WHO) (2015) or the National Institute on Aging (2019) suggest recommendations for healthy aging targeting the lifestyle of older adults. One common recommendation is "keep your mind and body active" to reduce the risk of developing diseases and disabilities that often occur with aging and to maintain autonomy and independence in older age. For example, engaging in intellectually activities can serve to buffer against cognitive decline (Hultsch, Hertzog, Small, & Dixon, 1999) and taking part in leisure-time physical activities like lap swimming or running may decrease the risk of a heart attack (Talbot, Morrell, Metter, & Fleg, 2002). Another common recommendation is "take charge of your health". This recommendation builds upon the assumption that individuals are able to make things happen proactively and intentionally by their own actions. For instance, older adults can pay attention to their weight and diet to influence changes in body mass index (BMI) and waist circumference, which are risk factors for many diseases. For instance, higher BMI and larger waist circumference are positively associated with coronary risk factors such as blood pressure or plasma lipids (Iwao et al., 2001). Moreover, certain personality traits may influence who takes charge of health and who does not (or does so to a lesser extent); it has been shown that conscientious individuals show better health outcomes

D. Aschwanden (✉)
Department of Geriatrics, College of Medicine, Florida State University, Tallahassee, FL, USA
e-mail: Damaris.Aschwanden@med.fsu.edu

M. Allemand
Department of Psychology and University Research Priority Program "Dynamics of Healthy Aging", University of Zurich, Zürich, Switzerland
e-mail: mathias.allemand@uzh.ch

P. L. Hill, M. Allemand (eds.), *Personality and Healthy Aging in Adulthood*, International Perspectives on Aging 26,
https://doi.org/10.1007/978-3-030-32053-9_10

through their action on social environmental factors, health-related behaviors, and psychophysiological mechanisms (Bogg & Roberts, 2004).

Recent research further supports these recommendations by showing that an active and self-directed life contributes to physical, social, and cognitive well-being (Gerstorf et al., 2016; Lathia, Sandstrom, Mascolo, & Rentfrow, 2017) and that older adults in fact desire an active, engaged, and healthy life (Huijg et al., 2017). But how do older individuals implement these recommendations in their daily lives? How do they translate the recommended strategies into action despite of daily obstacles and challenges? And why do some older individuals adhere to the recommended lifestyle modifications and factors associated with healthy aging, whereas others show low levels of adherence? Or more broadly, how should older adults be in motion and take action for healthy aging? Beyond that, what individual and situational characteristics may support an effective implementation of and compliance with strategies for an active and self-directed life? Whereas previous research investigated questions like these in the *long-term* (i.e., over longer time periods such as decades or years), little is known about them in the *short-term* (i.e., over shorter time periods such as days or weeks), including the dynamics, mechanisms and moderators of daily activities that contribute to healthy aging.

The present chapter proposes a conceptual framework that aims to explain how and why engaging in cognitive, physical, and social activities is associated with short-term healthy aging outcomes. The *Activities in Motion and in Action (AMA)* framework depicts an overview of the mechanisms that may underlie cognitive, physical, and social activities' successful short-term effects and the features that render these activities optimally effective. As an essential part of the framework, personality-informed interventions are suggested to promote the engagement in cognitive, physical and social activities of older adults (i.e., "be in motion and take action"). The first part of this chapter gives an overview about theoretical and empirical work on the long-term associations between personality, activities and healthy aging. The second part introduces the AMA framework and its different components such as engagement, outcomes, mechanisms, and moderators. The third section defines personality-informed interventions as well as the factors of such an intervention. The chapter concludes with research and practical implications.

10.1.1 Defining Cognitive, Physical and Social Activities

To keep older adults in motion and to encourage them to take action for healthy aging, stimulating the engagement of cognitive, physical and social activities is central. It is thus necessary to define these activities here. We define them in line with previous work (Christensen & MacKinnon, 1993; Hultsch, Hammer, & Small, 1993; Jopp & Hertzog, 2010). *Cognitive activities* include, for example, reading books, watching educational movies or a documentary, listening to the radio, or using a computer. *Physical activities* encompass participation in sports and recreation such as bicycling, hiking, aerobics, or tennis as well as physical work around

simple intentional positive activities, such as expressing gratitude or practicing kindness, leads to well-being and happiness. Whereas the positive-activity model focuses solely on positive activities related to long-term well-being and happiness, the AMA framework refers to a broader range of activities and their effects on short-term healthy aging outcomes. The AMA framework is shown in Fig. 10.1. It consists of five components: Engagement, outcomes, mechanisms, moderators and intervention. These components may influence each other, which is illustrated by five paths (see arrows 1–5 in Fig. 10.1). The component *intervention* is particularly relevant as it provides a starting point for future empirical studies. In the following, the different framework components are explained.

Engagement This component refers to the engagement of cognitive, physical and social activities. Stimulating activity engagement is central in the AMA framework, as it is assumed to be the recipe to keep older adults in motion and to encourage them to take action for healthy aging. Definitions of the activities were provided in the section before. Whereas the aforementioned definitions may refer to "general" or "typical" activity engagement, the AMA framework focuses on momentary activities in daily life.

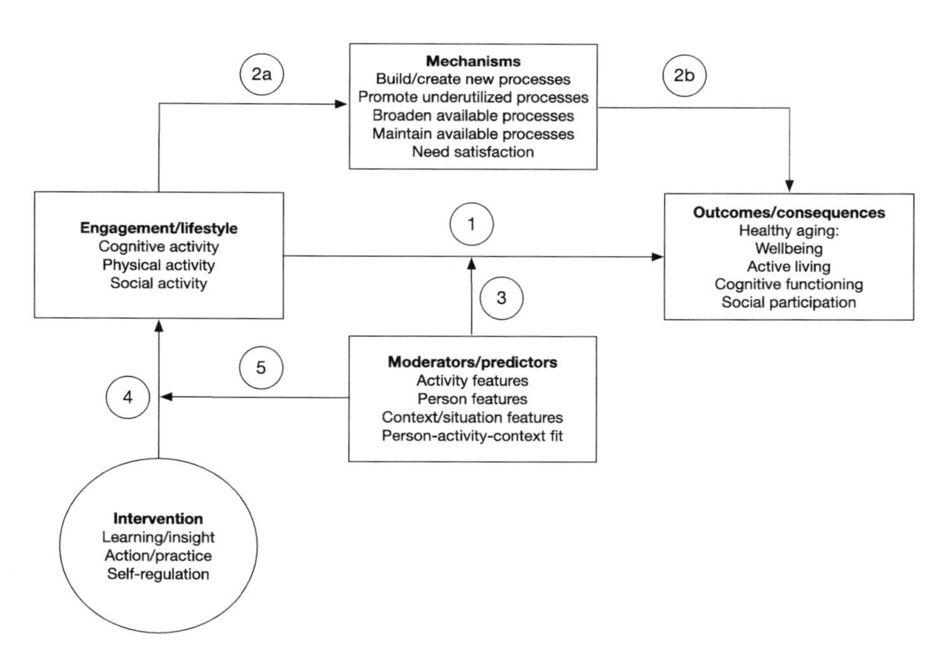

Fig. 10.1 The Activities in Motion and in Action (AMA) framework. The aim of the framework is to explain how and why the engagement in cognitive, physical and social activities ("be in motion and take action") is related to short-term and eventually long-term healthy aging outcomes, and how personality-informed interventions could target the engagement in these activities

Outcomes The AMA framework includes a broad range of short-term healthy aging outcomes such as well-being, active living, cognitive functioning and social participation. While a variety of definitions for healthy aging has been suggested, the AMA framework is based on the definition from the World Health Organization (WHO). The WHO defines *healthy aging* as "the process of developing and maintaining the functional ability that enables well-being in older age" (2015, p. 28). *Well-being* is broadly defined as "it includes happiness, satisfaction, and fulfilment" (p. 29). Adapted from the WHO's definition, the AMA framework defines *active living* as a way of life in which cognitive, physical and social activities are valued and integrated into daily living. *Cognitive functioning* can be defined as mental processes that lead to the acquisition of knowledge and describe the processes of thought (Kolb & Whishaw, 2006, p. 523). Finally, *social participation* may refer to the extent to which an individual participates in a broad range of social roles and relationships (Avison, McLeod, & Pescosolido, 2007).

Mechanisms The association between engagement and outcomes (see arrow 1 in Fig. 10.1) may be affected by different mechanisms (see arrows 2a and 2b in Fig. 10.1). The AMA framework proposes five broad mechanisms that may underlie successful short-term effects of engagement in the activities: (1) build or create new resources, (2) promote underutilized resources, (3) broaden available resources, (4) maintain available resources, and (5) satisfy basic psychological needs.

These mechanisms were derived from existing theories and previous long-term studies. For example, learning to acquire digital skills (i.e., cognitive activity) may *build* new resources in social participation, offering new opportunities to socialize and to participate in cultural, economic and political lives (cf. Hickman, Rogers, & Fisk, 2007). In the AMA framework, engaging in daily cognitive, physical, and social activities may *build and create new short-term resources* for an individual. For instance, if a lonely person begins to participate in social activities, social resources may be created and lead to social support in daily life.

Furthermore, engaging in leisure activities *promotes* a variety of social and physical resources (Coleman & Iso-Ahola, 1993; Iwasaki & Mannell, 2000; Iwasaki & Schneider, 2003). Leisure activities with others may provide social support and, in turn, mediate the stress-health relationship (Coleman & Iso-Ahola, 1993), enrich meaning of life (Carruthers & Hood, 2004), recovery from stress, and restoration of social and physical resources (Pressman et al., 2009), as well as help older adults adapt to potential restrictions of chronic conditions (Hutchinson & Nimrod, 2012) and overcome negative life events (e.g., losing a loved one) (Janke, Nimrod, & Kleiber, 2008). In terms of the AMA framework, engaging in daily cognitive, physical, and social activities may *promote underutilized resources*. This means, if someone used to play piano when young, the resumption of playing piano may promote underutilized cognitive resources and contribute to short-term cognitive functioning.

As another example, engaging in juggling has been shown to *broaden* neurobiological resources in the brain, leading to microscopic changes and macroscopic alterations and thus expanding the grey matter volume in specific brain areas

(Draganski et al., 2004). In the study by Draganski and colleagues, two groups - designated as jugglers and non-jugglers (both groups were inexperienced in juggling at the beginning) - were compared on region-specific grey and white matter in the brain. The juggler group learned a classic three-ball cascade juggling routine over 3 months. Group comparison showed significant regional differences in grey matter between jugglers and non-jugglers. This finding indicates that the structure of an adult human brain can alter in response to training, illustrating that learning-induced behavior may broaden available resources. Based on the assumption of the AMA framework, engaging in daily cognitive, physical, and social activities may *broaden available short-term resources*: If an individual regularly engages in cognitive activities that are an appropriate challenge for this person, his or her cognitive resources may be broadened and in turn contribute positively to short-term healthy aging outcomes.

Referring to the "use it or lose it" hypothesis (Hultsch et al., 1999), individuals who engage in activities that make significant loads on their cognitive skills will show greater *maintenance* of their cognitive resources (i.e., abilities) than individuals who are exposed to less complex environments with minimal cognitive loads. In turn, these cognitive resources may contribute positively to cognitive functioning. The AMA framework suggests that engaging in cognitive, physical, and social activities may help individuals to *maintain available resources*. For instance, if a person regularly plays tennis for fun, this may help to maintain physical and social resources, thus relating to both short-term objective and subjective health as well as social participation.

Finally, various *psychological needs* are commonly met by engaging in leisure activities (Newman, Tay, & Diener, 2014). For example, leisure activities that involve skill acquisition, such as learning a new language (cognitive activity) or how to play tennis (physical activity) afford opportunities for competence/mastery which in turn has been found to influence subjective well-being (Sonnentag & Fritz, 2007). Referring to the AMA framework, engaging in cognitive, physical, and social activities may boost well-being by *satisfying basic psychological needs*, such as autonomy (control), competence (efficacy), and relatedness (connectedness; Ryan & Deci, 2017). As such, an individual who engages in social activities may have an increased feeling of relatedness that in turn may contribute to increased well-being or greater social participation.

Moderators The association between engagement and outcomes (see arrow 1 in Fig. 10.1) may be affected by different moderators (see arrow 3 in Fig. 10.1). These moderators may not only influence the link between engagement and outcomes (see arrow 3 in Fig. 10.1), but also the intervention (see arrow 5 in Fig. 10.1). Similar to the positive-activity model (Lyubomirsky & Layous, 2013), the AMA framework proposes the following four features under which cognitive, physical, and social activities most effectively enhance short-term healthy aging outcomes: activity features, person features, context/situation features, and the person-activity fit.

First, activity features refer to characteristics of the activity, for example challenge, frequency, intensity, timing, and variety. Previous research showed that engaging in high-challenge activities (digital photography and quilting) increased modulation of brain activity in regions associated with attention and semantic processing, whereas engaging in low-challenge activities (socializing or performing low-challenge cognitive tasks) did not (McDonough, Haber, Bischof, & Park, 2015). Furthermore, participating in a variety of lifestyle activities appears more strongly associated with a reduced risk of incident cognitive impairment and dementia than participating more frequently in a single activity type (Carlson et al., 2012; Podewils et al., 2005; Wang et al., 2013). In the study of Wang et al. (2013), individuals who engaged in two or three activities even improved their cognitive performance, whereas those who engaged in one activity maintained their cognitive performance and those who did not participate in any activities experienced a global cognitive decline. A composite measure of the number of activities has further been associated with higher well-being (Menec, 2003), better physical functioning (Carlson et al., 2012), and lower risk of dementia (Karp et al., 2006). Thus, engaging in a variety of activity types (i.e., cognitive, physical, and social activities) is recommended by the AMA framework. Individuals who engage in a variety of activity types are more likely to tap into several resources (build/create, promote, broaden, maintain) and satisfy basic psychological needs (autonomy, competence, relatedness) that may lead to healthy aging outcomes.

Second, person features pertain to the individual practicing the activity (e.g., efforts, motivation, beliefs, personality). To benefit from the activities, individuals must effortfully engage in them (Lyubomirsky, Dickerhoof, Boehm, & Sheldon, 2011), be motivated (Deci & Ryan, 2000), and believe that their efforts will pay off (Bandura, 1986). In addition to people's motivation, efforts, and beliefs, people's personalities may affect how much they engage in a variety of activity types (Stephan et al., 2014). Stephan and colleagues found that individuals who scored higher on extraversion and openness were more likely to engage in a variety of activity types.

Third, context and situation features refer to the environment in which the individuals find themselves. For example, social network and social support can be considered as social environments which in turn affect social activities and healthy aging outcomes. This is exemplified by previous research showing that social environments influence cognitive performance and cognitive decline: Individuals who received more emotional support showed better baseline performance and better cognitive function at the 7.5-year follow-up, controlling for baseline cognitive function and several sociodemographic, behavioral, psychological, and health status predictors of cognitive aging (Seeman, Lusignolo, Albert, & Berkman, 2001). The AMA framework thus considers social environment as a crucial context variable supporting healthy aging.

Fourth, the person-activity-context fit is the match between activity, person, and context features. Although the above-mentioned features broadly influence those activities' success at enhancing short-term processes related to healthy aging, certain types of activities may be better for certain people given a certain context. The AMA framework assumes that activity, person, and context features interact with one another. The importance of this fit is supported by studies showing that the degree to which participants report enjoying an activity predicts how often they complete that activity (Schueller, 2010) and how much happiness they derive from it (Lyubomirsky, 2008).

10.2.1 Targeting Personality to Promote Engagement

To stimulate the regular engagement in cognitive, physical and social activities, the AMA framework suggests personalized interventions. To be more specific, the AMA framework considers individual differences in personality to personalize interventions, thus proposing personality-informed interventions (Chapman et al., 2014). This is important because a standard "one-size-fits-all" approach ignores individual differences in personality (Terracciano & Sutin, 2019). However, personality is crucial for health behaviors and health outcomes as highlighted previously in this chapter. In other words, individuals may have different reasons to participate in an intervention and may not as much benefit from a standard intervention as they could from a personality-informed intervention. Likewise, certain activities may work better for some individuals. It is thus time to shift the focus to the person to produce interventions that are more effective. If ultimately successful, such interventions are probably cheaper to develop than others such as medical treatments (Mroczek, 2014).

Existing personality-informed interventions have been successful in reducing depressive, anxiety, and conduct symptoms in high-risk youth (O'Leary-Barrett et al., 2013), reducing behavioral symptoms of dementia (Kolanowski, Litaker, Buettner, Moeller, & Costa, 2011), and increasing working memory performance (Studer-Luethi, Jaeggi, Buschkuehl, & Perrig, 2012). Therefore, the AMA framework recommends to tailor interventions that are compatible with an individual to facilitate adaptive behavior and activity change (Chapman et al., 2014; Robinson & Lachman, 2017). By considering individual differences in features of activity, person, context/situation and the person-activity fit, interventions can be established that are relevant to the unique goals or barriers of the individual. This personality-informed approach considers that there are types or clusters of people with commonalities who would respond to specific intervention elements (e.g., frequency or combination of activities) in similar ways. Thus, the goal is to tailor the interventions accordingly with the goal of sustained behavior and activity change (Robinson & Lachman, 2017). For example, if a physician suggests to an older patient that the patient should engage more often in cognitive activities, it is important to identify the specific barriers that might prevent the individual from doing so. If high neuroticism is indicated, then the intervention could start by targeting neuroticism. This could include specific cognitive-behavioral tasks (see Stieger et al., 2018, for an illustrative example of a personality intervention), along with providing the necessary skills and environmental supports to engage in more cognitive activity. But if the main barrier to becoming more cognitively engaged is lack of knowledge about how to do so rather than high neuroticism, this would suggest a focus on strategies to obtain the necessary information and skills. Given their implications towards healthy aging, providing personality-informed interventions to promote cognitive, physical and social activities should be a major public health concern.

Intervention The AMA framework represents a conceptual framework to derive personality-informed interventions and proposes that several intervention factors influence engagement (see arrow 4 in Fig. 10.1). One could also hypothesize that these factors affect mechanisms and moderators; however, this is beyond the scope of the present chapter. The following section focuses on general intervention factors as potential heuristic principles to promote that older adults are in motion and take action. Specifically, these factors are (1) action/practice, (2) learning/insight, and (3) self-regulation. The AMA framework focuses on these common factors rather than developing multiple intervention approaches that are unique to specific cognitive, physical and social interventions. This assumption is based on findings from psychotherapy research and social cognitive theory where outcomes can be largely explained by shared principles or common factors rather than by specific techniques or unique factors (Bandura, 1986; Lambert, 2013).

Learning/Insight This factor involves the individual to become aware of the motivational determinants such as expectations, motives, standards, goals, fears, beliefs, values, and wishes to engage in cognitive, physical and social activities. The intervention should thus stimulate learning (i.e., cognitive and emotional understanding of beliefs, expectations, or motives) and reflective processes (e.g., self-reflection). To realize learning and insight, it is necessary to primarily target people's cognitive-affective/reflective functioning. This can be achieved by carefully challenging the individual's basic assumptions, beliefs, expectations and motives, which in turn allow the individual to self-reflect, self-explore and self-narrate (cf. Allemand & Flückiger, 2017). This process may then result in new ideas and broader thoughts rather than sticking to old mindsets.

Action/Practice This factor refers to experiences that confer a better sense of self-efficacy and change in the individual's behaviors. Put differently, it relates to promoting adaptive (and changing maladaptive) cognitive, physical and social behaviors by learning new behaviors and skills. To promote these behaviors and achieve a change in daily life, it is necessary to identify appropriate behaviors, and practice them. Practice is focused on the person's concrete actions and behaviors (cf. Allemand & Flückiger, 2017).

Self-Regulation This factor refers to extrinsic and intrinsic processes responsible for monitoring, evaluating, and modifying behaviors. Self-efficacy plays a crucial role in this process, as it influences our thoughts, feelings, motivation, and action (Bandura, 1986). Hence, the intervention should target self-regulation as well as self-efficacy. On the one hand, the intervention's activities should stimulate self-regulation, for example, to overcome one's weaker self when it is raining outside, but one should go for a walk. On the other hand, the activities should support the individual's belief in his or her ability to achieve goals.

10.3 Implications

The AMA framework has great potential to inform future research on healthy aging. In contrast to past frameworks, it relies on short-term time periods such as days or weeks to better understand the complex and intertwined short-term dynamics of healthy aging. Further, it refers to a range of activities, includes broad mechanisms and moderators, and suggests intervention factors as potential heuristic principles. It also emphasizes the importance of personality-informed interventions to promote the engagement in cognitive, physical, and social activities. In aging and health psychology, we try to understand how older adults manage to maintain their health in interaction with their personalities, activities, abilities and their environmental contexts. Hence, it is essential to comprehend how the dynamics, mechanisms and moderators of daily activities are interrelated and contribute to healthy aging over shorter time periods such as days or weeks in addition to longer time periods such as years and decades. Today's technologies (e.g., smartphones, tablets) and the recent movements towards developing and designing personality-informed interventions allows us now to go one step further in this direction.

10.3.1 Research Implications

The AMA framework offers guidance for future empirical research efforts. One potential avenue for future research is to further accumulate evidence about the short-term links between personality and the engagement in cognitive, physical and social activities. For example, can measures of daily personality inform clinicians and interventionists regarding which older adults are more likely to engage in certain activities? More information on these links helps to refine personality-informed profiles and programs. It is important for future research to use multiple methods of assessment, such as observer ratings and partner-reports, mobile sensing data, objective activity measures, behavioral experiments, and daily life paradigms to eliminate the shared variance that occurs from using self-reports of personality and activity.

Future studies should develop and test personality-informed interventions to promote the engagement in cognitive, physical and social activities. To design personality-informed interventions, it is important to obtain a personality-informed profile. This means, participants need to complete a series of self-report questionnaires (e.g., personality, activities, resources, contexts, well-being, active living, social participation) and cognitive tasks (cognitive functioning). Furthermore, participants should report on their values, beliefs, goals and wishes and on what is important for them to be and to do what they have reason to value (cf. World Health Organization (WHO) (2015). Together, these indicators yield a personality-informed profile. Subsequently, an algorithm could identify individualized activities for each participant. Not only may the proportion of the activities differ, but also the activity itself within the same activity type. Generally speaking, the individualized activities should stimulate learning, practice and self-regulation. To tap into several resources and to satisfy basic psychological needs, the intervention should include a variety of activity types (cognitive, physical, and social). For the intervention phase, participants could be triggered to answer questions on states, behaviors, and short-term outcomes related to healthy aging (e.g., personality, activities, well-being, social participation, values, beliefs, goals, and cognitive functioning) on their mobile phones. At the end of the intervention phase, participants would fill in the same questionnaires completed at pre-intervention. The personality-informed intervention could be adapted and repeated for those participants yet to reach their personally important outcomes or goals.

Further research should use measurement-burst designs with the unique opportunity to study AMA processes and personality-informed interventions over different temporal intervals (e.g., Sliwinski, 2008). For example, it would be interesting to link short-term AMA relations with long-term healthy aging processes in order to examine research questions. Theories could be derived subsequently that contribute to the understanding of short-term dynamic processes and long-term development. In personality psychology, as an example, the TESSERA framework is such a theory that explicitly links short- and long-term processes of personality development by addressing different manifestations of personality and by being applicable to different personality characteristics (Wrzus & Roberts, 2017). The TESSERA framework states that long-term personality development occurs due to repeated short-term processes, while these short-term processes are recursive sequences of triggering situations, expectancy, states/state expressions, and reactions (TESSERA). Likewise, one can hypothesize that long-term healthy aging occurs because of daily repeated cognitive, physical, and social activities.

10.3.2 Practical Implications

Finally, the AMA framework also has great potential to inform (1) service providers that offer advice, assistance and organize leisure time activities for older adults, and (2) public health and aging organizations that provide recommendations for healthy aging. For example, knowing that (a) personality is associated with the engagement in cognitive, physical and social activities and that (b) activity, person, and context features interact with one another might be helpful to provide more differentiated recommendations. For instance, if the goal is to keep older adults mentally and physically active, the recommendations for conscientious and neurotic people may look different. Whereas a simple recommendation such as "keep your mind and body active" may be enough for conscientious individuals, it might be improved for neurotic people by offering the opportunity to test for cognitive and physical status, pointing out the benefits of regular engagement in these activities, and regular reminders provided by physicians or a mobile application.

Furthermore, considering that (c) individuals have different reasons to participate in an intervention and (d) factors such as action/practice, learning/insight, and self-regulation are associated with intervention outcomes might be useful for intervention studies. Matching individuals to the intervention that is relevant to their unique goals and stimulates action/practice, learning/insight, and self-regulation means that the money spent dispensing the intervention will lead to greater net benefit and cost effectiveness. Moreover, interventions might be rendered more effective by administering them only to persons who are more likely to respond, rather than to all comers (i.e., adapting personality-informed interventions). The challenge is now to test the AMA framework and subsequently translate its findings into public health practice.

10.4 Conclusion

The AMA framework represents a unique framework of short-term dynamics and underlying processes of healthy aging that typically cannot be covered (a) in studies conducted in the laboratory (Bolger, Davis, & Rafaeli, 2003) and (b) as they occur alongside long-term developmental processes (Diehl, Hooker, & Sliwinski, 2015). It is also one of the first attempts that aims to summarize and conceptualize the multifactorial, intertwined, and complex etiology of healthy aging. It provides support to study research questions revolving around healthy aging in daily life. The goal of the Activities in Motion and in Action (AMA) framework is to elucidate how and why older adults are in motion and take action for healthy aging. Based on this framework, personality-informed interventions offer the possibility to examine (1) how older individuals engage in cognitive, physical and social activities (i.e., *are in motion and take action*) and what they learn from the experiences related to these activities (*insight*); (2) how older individuals deal with opportunities and challenges to engage in activities (*self-regulation*); (3) how the activity, person, context/situation features, and the person-activity fit influence the engagement in activities (*moderators*); and (4) how engagement in activities promotes individual well-being and subjective health (*short-term healthy outcomes*). This conceptual framework aims to stimulate future conceptual/theoretical advances and empirical personality-informed intervention studies to promote engagement in cognitive, physical and social activities.

References

Adams, K. B., Leibbrandt, S., & Moon, H. (2011). A critical review of the literature on social and leisure activity and wellbeing in later life. *Ageing and Society, 31*, 683–712. https://doi.org/10.1017/S0144686X10001091

Allemand, M., & Flückiger, C. (2017). Changing personality traits: Some considerations from psychotherapy process-outcome research for intervention efforts on intentional personality change. *Journal of Psychotherapy Integration, 27*, 476–494. https://doi.org/10.1037/int0000094

Avison, W. R., McLeod, J. D., & Pescosolido, B. A. (Eds.). (2007). *Mental health, social mirror*. New York, NY: Springer.

Bandura, A. (1986). *Social foundations of thought and action: A social cognitive theory*. Englewood Cliffs, NJ: Prentice Hall.

Bielak, A. A. M., Anstey, K. J., Christensen, H., & Windsor, T. D. (2012). Activity engagement is related to level, but not change in cognitive ability across adulthood. *Psychology and Aging, 27*, 219–228. https://doi.org/10.1037/a0024667

Bogg, T., & Roberts, B. W. (2004). Conscientiousness and health-related behaviors: A meta-analysis of the leading behavioral contributors to mortality. *Psychological Bulletin, 130*, 887–919. https://doi.org/10.1037/0033-2909.130.6.887

Bolger, N., Davis, A., & Rafaeli, E. (2003). Diary methods: Capturing life as it is lived. *Annual Review of Psychology, 54*, 579–616. https://doi.org/10.1146/annurev.psych.54.101601.145030

Carlson, M. C., Parisi, J. M., Xia, J., Xue, Q.-L., Rebok, G. W., Bandeen-Roche, K., & Fried, L. P. (2012). Lifestyle activities and memory: Variety may be the spice of life. The Women's Health and Aging Study II. *Journal of the International Neuropsychological Society, 18*, 286–294. https://doi.org/10.1017/S135561771100169X

Carruthers, C. P., & Hood, C. D. (2004). The power of the positive: Leisure and Well-being. *Therapeutic Recreation Journal, 38*, 225–245.

Cattan, M., White, M., Bond, J., & Learmouth, A. (2005). Preventing social isolation and loneliness among older people: A systematic review of health promotion interventions. *Ageing and Society, 25*, 41–67. https://doi.org/10.1017/S0144686X04002594

Chapman, B. P., Hampson, S., & Clarkin, J. (2014). Personality-informed interventions for healthy aging: Conclusions from a National Institute on Aging work group. *Developmental Psychology, 50*, 1426–1441. https://doi.org/10.1037/a0034135

Chase, J.-A. D. (2013). Physical activity interventions among older adults: A literature review. *Research and Theory for Nursing Practice, 27*, 53–80.

Christensen, H., & MacKinnon, A. (1993). The association between mental, social and physical activity and cognitive performance in young and old subjects. *Age and Ageing, 22*, 175–182. https://doi.org/10.1093/ageing/22.3.175

Coleman, D., & Iso-Ahola, S. E. (1993). Leisure and health: The role of social support and self-determination. *Journal of Leisure Research, 25*, 111–128. https://doi.org/10.1080/00222216.1993.11969913

Davis, M. G., & Fox, K. R. (2007). Physical activity patterns assessed by accelerometry in older people. *European Journal of Applied Physiology, 100*, 581–589. https://doi.org/10.1007/s00421-006-0320-8

Deci, E. L., & Ryan, R. M. (2000). Self-determination theory and the facilitation of intrinsic motivation, social development, and wellbeing. *American Psychologist, 55*, 68–78. https://doi.org/10.1037/0003-066X.55.1.68

Diehl, M., Hooker, K., & Sliwinski, M. J. (Eds.). (2015). *Handbook of intraindividual variability across the life-span*. New York, NY: Routledge, Taylor & Francis Group.

Draganski, B., Gaser, C., Busch, V., Schuierer, G., Bogdahn, U., & May, A. (2004). Changes in grey matter induced by training: Neuroplasticity. *Nature, 427*, 311–312. https://doi.org/10.1038/427311a

Gerstorf, D., Hoppmann, C. A., Löckenhoff, C. E., Infurna, F. J., Schupp, J., Wagner, G. G., & Ram, N. (2016). Terminal decline in well-being: The role of social orientation. *Psychology and Aging, 31*, 149–165. https://doi.org/10.1037/pag0000072

Glass, T. A., de Leon, C. M., Marottoli, R. A., & Berkman, L. F. (1999). Population based study of social and productive activities as predictors of survival among elderly Americans. *BMJ, 319*, 478–483. https://doi.org/10.1136/bmj.319.7208.478

Hickman, J. M., Rogers, W. A., & Fisk, A. D. (2007). Training older adults to use new technology. *The Journals of Gerontology. Series B, Psychological Sciences and Social Sciences, 62*, 77–84. https://doi.org/10.1093/geronb/62.special_issue_1.77

Hill, P. L., Turiano, N. A., Hurd, M. D., Mroczek, D. K., & Roberts, B. W. (2011). Conscientiousness and longevity: An examination of possible mediators. *Health Psychology, 30*, 536–541. https://doi.org/10.1037/a0023859

Hogan, M. J., Staff, R. T., Bunting, B. P., Deary, I. J., & Whalley, L. J. (2012). Openness to experience and activity engagement facilitate the maintenance of verbal ability in older adults. *Psychology and Aging, 27*, 849–854. https://doi.org/10.1037/a0029066

Huijg, J. M., van Delden, A., van der Ouderaa, F. J., Westendorp, R. G., Slaets, J. P., & Lindenberg, J. (2017). Being active, engaged, and healthy: Older persons' plans and wishes to age successfully. *The Journals of Gerontology Series B: Psychological Sciences and Social Sciences, 72*, 228–236. https://doi.org/10.1093/geronb/gbw107

Hultsch, D. F., Hammer, M., & Small, B. J. (1993). Age differences in cognitive performance in later life: Relationships to self-reported health and activity life style. *Journal of Gerontology, 48*, 1–11. https://doi.org/10.1093/geronj/48.1.P1

Hultsch, D. F., Hertzog, C., Small, B. J., & Dixon, R. A. (1999). Use it or lose it: Engaged lifestyle as a buffer of cognitive decline in aging? *Psychology and Aging, 14*, 245–263. https://doi.org/10.1037/0882-7974.14.2.245

Hutchinson, S. L., & Nimrod, G. (2012). Leisure as a resource for successful aging by older adults with chronic health conditions. *The International Journal of Aging and Human Development, 74*, 41–65. https://doi.org/10.2190/AG.74.1.c

Iwao, S., Iwao, N., Muller, D. C., Elahi, D., Shimokata, H., & Andres, R. (2001). Does waist circumference add to the predictive power of the body mass index for coronary risk? *Obesity Research, 9*, 685–695. https://doi.org/10.1038/oby.2001.93

Iwasaki, Y., & Mannell, R. C. (2000). Hierarchical dimensions of leisure stress coping. *Leisure Sciences, 22*, 163–181. https://doi.org/10.1080/01490409950121843

Iwasaki, Y., & Schneider, I. E. (2003). Leisure, stress, and coping: An evolving area of inquiry. *Leisure Sciences, 25*, 107–113. https://doi.org/10.1080/01490400306567

Janke, M. C., Nimrod, G., & Kleiber, D. A. (2008). Reduction in leisure activity and well-being during the transition to widowhood. *Journal of Women & Aging, 20*, 83–98. https://doi.org/10.1300/J074v20n01_07

John, O. P., Naumann, L. P., & Soto, C. J. (2008). Paradigm shift to the integrative big five trait taxonomy: History, measurement, and conceptual issues. In O. P. John, R. W. Robins, & L. A. Pervin (Eds.), *Handbook of personality: Theory and research* (3rd ed., pp. 114–158). New York, NY: Guilford.

Jopp, D., & Hertzog, C. (2007). Activities, self-referent memory beliefs and cognitive performance: Evidence for direct and mediated relations. *Psychology and Aging, 22*, 811–825. https://doi.org/10.1037/0882-7974.22.4.811

Jopp, D. S., & Hertzog, C. (2010). Assessing adult leisure activities: An extension of a self-report activity questionnaire. *Psychological Assessment, 22*, 108–120. https://doi.org/10.1037/a0017662

Karp, A., Paillard-Borg, S., Wang, H., Silverstein, M., Winblad, B., & Fratiglioni, L. (2006). Mental, physical and social components in leisure activities equally contribute to decrease dementia risk. *Dementia and Geriatric Cognitive Disorders, 21*, 65–73. https://doi.org/10.1159/000089919

Kolanowski, A., Litaker, M., Buettner, L., Moeller, J., & Costa, P. T., Jr. (2011). A randomized clinical trial of theory-based activities for the behavioral symptoms of dementia in nursing home residents. *Journal of the American Geriatrics Society, 59*, 1032–1041. https://doi.org/10.1111/j.1532-5415.2011.03449.x

Kolb, B., & Whishaw, I. Q. (2006). *An introduction to brain and behavior* (2nd ed.). New York, NY: Worth Publishers.

Lambert, M. J. (2013). Outcome in psychotherapy: The past and important advances. *Psychotherapy, 50*, 42–51. https://doi.org/10.1037/a0030682

Lathia, N., Sandstrom, G. M., Mascolo, C., & Rentfrow, P. J. (2017). Happier people live more active lives: Using smartphones to link happiness and physical activity. *PLoS ONE, 12*, 1–13. https://doi.org/10.1371/journal.pone.0160589

Lyubomirsky, S. (2008). *The how of happiness: A scientific approach to getting the life you want*. New York, NY: Penguin Press.

Lyubomirsky, S., Dickerhoof, R., Boehm, J. K., & Sheldon, K. M. (2011). Becoming happier takes both a will and a proper way: An experimental longitudinal intervention to boost well-being. *Emotion, 11*, 391–402. https://doi.org/10.1037/a0022575

Lyubomirsky, S., & Layous, K. (2013). How do simple positive activities increase Well-being? *Current Directions in Psychological Science, 22*, 57–62. https://doi.org/10.1177/0963721412469809

Martin, M., Clare, L., Altgassen, A. M., Cameron, M. H., & Zehnder, F. (2011). Cognition-based interventions for healthy older people and people with mild cognitive impairment. In the Cochrane Collaboration (Eds.), *Cochrane database of systematic reviews*. Chichester, UK: Wiley.

Mazzucchelli, T. G., Kane, R. T., & Rees, C. S. (2010). Behavioral activation interventions for Well-being: A meta-analysis. *The Journal of Positive Psychology, 5*, 105–121. https://doi.org/10.1080/17439760903569154

McCrae, R. R., & Costa, P. T., Jr. (2008). A five-factor theory of personality. In L. A. Pervin & O. P. John (Eds.), *Handbook of personality: Theory and research* (3rd ed., pp. 159–181). New York, NY: Guilford.

McDonough, I. M., Haber, S., Bischof, G. N., & Park, D. C. (2015). The synapse project: Engagement in mentally challenging activities enhances neural efficiency. *Restorative Neurology and Neuroscience, 33*, 865–882. https://doi.org/10.3233/RNN-150533

Menec, V. H. (2003). The relation between everyday activities and successful aging: A 6-year longitudinal study. *The Journals of Gerontology Series B: Psychological Sciences and Social Sciences, 58*, 74–82. https://doi.org/10.1093/geronb/58.2.S74

Mroczek, D. K. (2014). Personality plasticity, healthy aging, and interventions. *Developmental Psychology, 50*, 1470–1474. https://doi.org/10.1037/a0036028

National Institute on Aging. (2019). *What do we know about healthy aging?* Retrieved April 1, 2019, from https://www.nia.nih.gov/health/what-do-we-know-about-healthy-aging

Newman, D. B., Tay, L., & Diener, E. (2014). Leisure and subjective well-being: A model of psychological mechanisms as mediating factors. *Journal of Happiness Studies, 15*, 555–578. https://doi.org/10.1007/s10902-013-9435-x

O'Leary-Barrett, M., Topper, L., Al-Khudhairy, N., Pihl, R. O., Castellanos-Ryan, N., Mackie, C. J., & Conrod, P. J. (2013). Two-year impact of personality-targeted, teacher-delivered interventions on youth internalizing and externalizing problems: A cluster-randomized trial. *Journal of the American Academy of Child & Adolescent Psychiatry, 52*, 911–920. https://doi.org/10.1016/j.jaac.2013.05.020

Paillard-Borg, S., Wang, H.-X., Winblad, B., & Fratiglioni, L. (2009). Pattern of participation in leisure activities among older people in relation to their health conditions and contextual factors: A survey in a Swedish urban area. *Ageing and Society, 29*, 803–821. https://doi.org/10.1017/S0144686X08008337

Podewils, L., Guallar, E., Kuller, L., Fried, L., Lopez, O., Carlson, M., & Lyketsos, C. (2005). Physical activity, APOE genotype, and dementia risk: Findings from the cardiovascular health cognition study. *American Journal of Epidemiology, 161*, 639–651. https://doi.org/10.1093/aje/kwi092

Pressman, S. D., Matthews, K. A., Cohen, S., Martire, L. M., Scheier, M., Baum, A., & Schulz, R. (2009). Association of enjoyable leisure activities with psychological and physical well-being. *Psychosomatic Medicine, 71*, 725–732. https://doi.org/10.1097/PSY.0b013e3181ad7978

Rhodes, R. E., & Smith, N. E. I. (2006). Personality correlates of physical activity: A review and meta-analysis. *British Journal of Sports Medicine, 40*, 958–965. https://doi.org/10.1136/bjsm.2006.028860

Roberts, B. W., Kuncel, N. R., Shiner, R., Caspi, A., & Goldberg, L. R. (2007). The power of personality: The comparative validity of personality traits, socioeconomic status, and cognitive ability for predicting important life outcomes. *Perspectives on Psychological Science, 2*, 313–345. https://doi.org/10.1111/j.1745-6916.2007.0047

Robinson, S. A., & Lachman, M. E. (2017). Perceived control and aging: A mini-review and directions for future research. *Gerontology, 63*, 435–442. https://doi.org/10.1159/000468540

Rowe, J. R., & Kahn, R. L. (1998). *Successful aging*. New York, NY: Pantheon Books.

Ryan, R. M., & Deci, E. L. (2017). *Self-determination theory: Basic psychological needs in motivation, development, and wellness*. New York, NY: Guilford Press.

Ryu, J., & Heo, J. (2018). Relationships between leisure activity types and well-being in older adults. *Leisure Studies, 37*, 331–342. https://doi.org/10.1080/02614367.2017.1370007

Schneider, N., & Yvon, C. (2013). A review of multidomain interventions to support healthy cognitive ageing. *The Journal of Nutrition, Health & Aging, 17*, 252–257. https://doi.org/10.1007/s12603-012-0402-8

Schueller, S. M. (2010). Preferences for positive psychology exercises. *The Journal of Positive Psychology, 5*, 192–203. https://doi.org/10.1080/17439761003790948

Seeman, T. E., Lusignolo, T. M., Albert, M., & Berkman, L. (2001). Social relationships, social support, and patterns of cognitive aging in healthy, high-functioning older adults: MacArthur studies of successful aging. *Health Psychology, 20*, 243–255. https://doi.org/10.1037/0278-6133.20.4.243

Shaw, B. A., Liang, J., Krause, N., Gallant, M., & McGeever, K. (2010). Age differences and social stratification in the long-term trajectories of leisure-time physical activity. *The Journals of Gerontology Series B: Psychological Sciences and Social Sciences, 65*, 756–766. https://doi.org/10.1093/geronb/gbq073

Sliwinski, M. J. (2008). Measurement-burst designs for social health research. *Social and Personality Psychology Compass, 2*, 245–261. https://doi.org/10.1111/j.1751-9004.2007.00043.x

Sonnentag, S., & Fritz, C. (2007). The recovery experience questionnaire: Development and validation of a measure for assessing recuperation and unwinding from work. *Journal of Occupational Health Psychology, 12*, 204–221. https://doi.org/10.1037/1076-8998.12.3.204

Stephan, Y., Boiché, J., Canada, B., & Terracciano, A. (2014). Association of personality with physical, social, and mental activities across the lifespan: Findings from US and French samples. *British Journal of Psychology, 105*, 564–580. https://doi.org/10.1111/bjop.12056

Stieger, M., Nissen, M., Rüegger, D., Kowatsch, T., Flückiger, C., & Allemand, M. (2018). PEACH, a smartphone- and conversational agent-based coaching intervention for intentional personality change: Study protocol of a randomized, wait-list controlled trial. *BMC Psychology, 6*. https://doi.org/10.1186/s40359-018-0257-9

Strickhouser, J. E., Zell, E., & Krizan, Z. (2017). Does personality predict health and well-being? A metasynthesis. *Health Psychology, 36*, 797–810. https://doi.org/10.1037/hea0000475

Studer-Luethi, B., Jaeggi, S. M., Buschkuehl, M., & Perrig, W. J. (2012). Influence of neuroticism and conscientiousness on working memory training outcome. *Personality and Individual Differences, 53*, 44–49. https://doi.org/10.1016/j.paid.2012.02.012

Talbot, L. A., Morrell, C. H., Metter, E. J., & Fleg, J. L. (2002). Comparison of cardiorespiratory fitness versus leisure time physical activity as predictors of coronary events in men aged < or = 65 years and > 65 years. *The American Journal of Cardiology, 89*, 1187–1192. https://doi.org/10.1016/S0002-9149(02)02302-0

Terracciano, A., & Sutin, A. R. (2019). Personality and Alzheimer's disease: An integrative review. *Personality Disorders, Theory, Research, and Treatment, 10*, 4–12. https://doi.org/10.1037/per0000268

Toman, J., Klímová, B., & Vališ, M. (2018). Multidomain lifestyle intervention strategies for the delay of cognitive impairment in healthy aging. *Nutrients, 10*, 1–10. https://doi.org/10.3390/nu10101560

Wang, H. X., Jin, Y., Hendrie, H. C., Liang, C., Yang, L., Cheng, Y., … Gao, S. (2013). Late life leisure activities and risk of cognitive decline. *The Journals of Gerontology Series A: Biological Sciences and Medical Sciences, 68*, 205–213. https://doi.org/10.1093/gerona/gls153

Williams, K. N., & Kemper, S. (2010). Interventions to reduce cognitive decline in aging. *Journal of Psychosocial Nursing and Mental Health Services, 48*, 42–51. https://doi.org/10.3928/02793695-20100331-03

Wilson, R. S., Krueger, K. R., Gu, L., Bienias, J. L., Mendes de Leon, C. F., & Evans, D. A. (2005). Neuroticism, extraversion, and mortality in a defined population of older persons. *Psychosomatic Medicine, 67*, 841–845. https://doi.org/10.1097/01.psy.0000190615.20656.83

World Health Organization (WHO). (2015). *World report on Ageing and Health*. Retrieved June 3, 2018, from http://apps.who.int/iris/bitstream/10665/186463/1/9789240694811_eng.pdf?ua=1

Wrzus, C., & Roberts, B. W. (2017). Processes of personality development in adulthood: The Tessera framework. *Personality and Social Psychology Review, 21*, 253–277. https://doi.org/10.1177/1088868316652279

Yaffe, K., Fiocco, A. J., Lindquist, K., Vittinghoff, E., Simonsick, E. M., Newman, A. B., … Harris, T. B. (2009). Predictors of maintaining cognitive function in older adults: The health ABC study. *Neurology, 72*, 2029–2035. https://doi.org/10.1212/WNL.0b013e3181a92c36

Chapter 11
Personality and Cognitive Health in Aging

Brennan R. Payne and Monika Lohani

11.1 Introduction

The world is getting older. Both the absolute number of older adults and the relative proportion of older adults in the population is increasing dramatically around the entire world (World Health Organization, 2015). Because of this, the age distribution across much of the world has shifted away from the classic 'population pyramid'—with a larger concentration of younger adults and a diminishing population of older adults—to a 'population rectangle' reflecting an increasingly growing segment of the population aged 65 and older. With this aging society comes growing concerns about the effects of senescence on cognitive and brain health. Indeed, adults in midlife, as young as their 40's, hold serious concerns about cognitive functioning, and many hold quite pessimistic views about maintaining brain health into old age (e.g., Vaportzis & Gow, 2018). Although one of the primary health concerns of older adults' centers on the development of Alzheimer's disease (AD) and related dementias, even in the absence of any neurodegenerative diseases, many aspects of cognitive functioning show normative changes in healthy adult development.

The normative senescence process brings about diffuse changes in many dimensions of brain and cognitive functioning in healthy aging. One of the most robust findings in the cognitive aging literature is the demonstration of negative age-related trajectories for cognitive abilities, including episodic memory (Park et al., 1996), working memory (Bopp & Verhaeghen, 2005), processing speed (Eckert, 2011; Salthouse, 1996; Salthouse & Madden, 2013), reasoning (Schaie, 1996), and

B. R. Payne (✉)
Department of Psychology, Neuroscience Program, and Center on Aging, University of Utah, Salt Lake City, UT, USA
e-mail: brennan.payne@utah.edu

M. Lohani
Department of Educational Psychology, University of Utah, Salt Lake City, UT, USA

P. L. Hill, M. Allemand (eds.), *Personality and Healthy Aging in Adulthood*, International Perspectives on Aging 26,
https://doi.org/10.1007/978-3-030-32053-9_11

executive control (Braver & West, 2008). Indeed, a general picture of normal cognitive aging is marked by decline in this constellation of highly-interrelated cognitive abilities, often termed "mental mechanics" or "fluid" cognitive abilities, those that require the ability to maintain and quickly transform information and effectively control attention (Deary, Penke, & Johnson, 2010). At the same time, not all cognitive functions change the same with advancing age. In other domains, a maintenance of functioning or even improvement, is observed in so-called "crystallized" intelligence, including world knowledge and verbal semantic memory (Salthouse & Ferrer-Caja, 2003). With advancing age, literate adults show preserved long-term memory for verbal information, including scoring upwards of 1 standard deviation above their younger counterparts in vocabulary assessments (Verhaeghen, 2003), as well as showing evidence for preserved semantic priming, reflecting efficient retrieval of lexical information in aging (Payne, Gao, Noh, Anderson, & Stine-Morrow, 2012; Federmeier, Van Petten, Schwartz, & Kutas, 2003; Lien et al., 2006). This divergent set of developmental trajectories, in which fluid abilities show negative age-related trajectories while crystallized abilities are maintained or grow, is one of the most well-replicated patterns of developmental change, observed both cross-sectionally and longitudinally (e.g., Salthouse, 2019).

Within these average developmental trajectories of change, there exist large individual differences in trajectories of cognitive change such that some individuals are able to maintain high levels of cognitive functioning until late in life while others show accelerated and abnormal rates of decline (Hertzog, Kramer, Wilson, & Lindenberger, 2008; Salthouse, 1996, 2019; Salthouse & Ferrer-Caja, 2003; Salthouse & Madden, 2013). Moreover, there exists substantial inter-individual variability in risk for the development of *non-normative* trajectories of cognitive aging, a term we will use throughout to broadly refer to both the development of probable dementia or clinically-relevant cognitive impairment. For instance, a growing literature shows that individuals vary considerably in the degree to which accumulating AD neuropathology manifests itself in terms of functional impairment in cognitive performance. This has led researchers to the argue that some individuals have greater neural or cognitive "reserve" capacity than others (for reviews see e.g., Manly, Touradji, Tang, & Stern, 2003; Stern, 2012; Stine-Morrow, Parisi, Morrow, & Park, 2008). Indeed, a substantial focus of the cognitive aging literature has been to identify the risk and protective factors that may be responsible for this considerable heterogeneity in trajectories of cognitive function in aging, with the goal of harnessing this information to create interventions to promote cognitive and brain health (see Hertzog et al., 2008; Smith, 2016). What are these pathways for cognitive enrichment in the face of normative decline? Several factors such as cardiovascular health, diet, fitness, social engagement, literacy ability, educational attainment, adult occupational complexity, and lifestyle have all appeared as important protective factors underlying maintenance of cognitive function and increased reserve capacity in older adulthood (Smith, 2016).

Important in this discussion then is understanding the mechanisms that can promote such healthy and engaging lifestyles in old age. Given that personality plays a role in lifestyle behaviors, discussions regarding the potential that personality traits and dispositional characteristics may play in shaping trajectories of cognitive aging are growing. Recently, Hill and Payne (2017) outlined the need for researchers to more thoughtfully consider the role of personality, attitudes, beliefs, and other dispositional characteristics in research on cognitive aging and dementia prevention. Hill and Payne (2017) conceptualized of personality as a (1) *predictor* of concurrent cognitive ability in older adulthood, cognitive decline, and risk for cognitive impairment in aging, (2) as a *precursor* to cognitive health risks in aging, and (3) as a *promoter* of (cognitive) intervention responsiveness and adherence. We aim to expand upon this outline, providing further justification and evidence highlighting the 'three P's' underlying the relationships between personality and cognitive aging.

11.2 Personality as a Predictor of Cognitive Health in Aging

11.2.1 Personality-Cognitive Ability Correlations

A growing research base has now established consistent relationships between personality traits and individual differences in concurrent cognitive functioning and subsequent cognitive change in aging (Sutin et al., 2019). Some of the first studies examining concurrent personality-cognition relationships focused on relationships between personality and gross-scale measures of aptitude and fluid intelligence in young adults (Ackerman & Heggestad, 1997; Lorge, 1940). Much of this early work was motivated by the desire to show that dimensions of personality assessments were not completely reducible to measures of intellectual ability or cognitive function, but rather that measures of personality, cognitive functioning, and intellectual interests formed partially overlapping 'trait complexes' (Ackerman & Heggestad, 1997). In Lorge's (1940) first review, median effect sizes between measures of intellectual functioning and personality traits were $r = .04$, with effects ranging from +.079 to −.049 across assessments. Similarly, in a meta-analysis of over 100 studies and 2000 correlations in non-clinical samples, Ackerman and Heggestad (1997) found small (effects rarely exceeded $r = .20$) but statistically significant correlations between ability and personality for about half of the relationships they examined. Despite the modest effect sizes, two clear trends emerged. First, there were consistent negative associations between traits associated with neuroticism and intellectual ability. Second, positive relationships were more consistently observed for openness and intellectual engagement, particularly for abilities loading on crystallized intelligence and knowledge. Following these initial meta-analyses, more systematic examinations of personality and cognition relationships have been explored.

Schaie, Willis, and Caskie (2004) provided one of the first systematic examinations of concurrent and longitudinal relationships between cognition and personality in older adults, in a cohort of over 2000 adults from the Seattle Longitudinal Study. Measures of reasoning, spatial cognition, psychomotor speed, numeracy, verbal ability, and memory were examined as a function of individual differences in a wide range of personality traits, including Big Five traits as well as a broader 13-dimensional factor model derived from the Test of Behavioral Rigidity, including traits such as *honesty, interest in science,* and *conservativism.* Similar to Ackerman and Heggestad's meta-analysis, results showed small-to-modest relationship between personality and cognitive function, with the exception of openness, which showed larger effect sizes (r = .3–.4 range) with measures of fluid and crystallized ability. Similarly, Soubelet and Salthouse (2011) found consistent relationships between openness and measures of reasoning and verbal episodic memory in a large sample of over 2000 adults ranging in age between 18–96 years old. Interestingly, they found that personality-cognition relationships were largely similar across age-range, with similar effect sizes in younger, middle-aged, and older adults. Moreover, higher neuroticism emerged as a small but reliable negative predictor of cognitive functioning. Perhaps unsurprisingly then, given the shared relationships between trait openness and intellect, higher endorsement of openness to experiences is associated with better cognitive performance across the lifespan. Although findings with conscientiousness have been less consistent than those of openness and neuroticism, findings do suggest that higher conscientiousness is weakly associated with better cognitive functioning.

Part of this inconsistency in the literature may be due in part to differences in the factors leading to relationships between cognitive functioning and trait conscientiousness. For example, some evidence suggests that conscientiousness is positively related to cognitive abilities because it influences health behaviors, which, in turn, are protective against age–related declines in brain and cognitive functioning (e.g., Sutin et al., 2011, 2019). On the other hand, conscientiousness has been negatively linked to cognitive functions, potentially because individuals with lower cognitive abilities may become more hardworking and organized over time in order to compensate for their lower functionality (Chamorro-Premuzic, Furnham, & Moutafi, 2004; Moutafi, Furnham, & Crump, 2003; Moutafi, Furnham, & Paltiel, 2004; Rammstedt, Danner, & Martin, 2016; c.f. Murray, Johnson, McGue, & Iacono, 2014).

In contrast, effects of extraversion and agreeableness have been less consistently found to relate to cognitive abilities (Curtis, Windsor, & Soubelet, 2015). Although these effect sizes tend to be modest across studies, an excellent meta-analysis by Luchetti, Terracciano, Stephan, and Sutin (2016) showed that these associations were generally larger than that of many risk factors currently well-understood to impact cognitive health in aging– including hypertension, diabetes, obesity, smoking, and physical inactivity. These findings suggest that there are reliable relationships between personality and cognitive health and aging.

11.2.2 Personality and Cognitive Impairment in Aging

A companion literature has begun exploring the role that personality plays in non-normative cognitive aging, including in dementia and mild cognitive impairment. The term dementia describes a cluster of symptoms associated with cognitive deficits that are severe enough to impact everyday activities of daily living. Alzheimer's dementia, the most commonly diagnosed dementia, is a progressive neurodegenerative disease comprising of severe deficits in multiple cognitive domains, including memory, reasoning, language processing and production, and executive functioning, in addition to emotional and behavioral symptoms. Mild cognitive impairment (MCI) is a transient pre-demented state that occurs between normal and pathological cognitive aging, marked by a concern regarding a change in cognition, an impairment in one or more cognitive domains, but a relative preservation of independence of functional abilities.

A meta-analysis by Terracciano et al. (2014) found evidence for consistent relationships between Neuroticism, Conscientiousness, and risk for Alzheimer's dementia. Individuals in the top quartile of Neuroticism or the lowest quartile of Conscientiousness had a 3-fold increased risk of incident AD. Conscientiousness may hold particular importance for predicting patterns of cognitive aging, because conscientious individuals may engage in more memory recall and long-term planning, to maintain their dispositionally organized and self-controlled manner. There were also weaker but significant effects for openness and agreeableness. This literature suggests that personality may have unique predictive capacity for clinically significant cognitive impairment over and above cross-sectional associations with cognitive performance. As such, recent work has begun to focus on relationships between personality and risk factors associated with dementia and MCI development in aging (for a review see Segerstrom, 2018 and Terracciano & Sutin, 2019).

In previous work, Payne and Stine-Morrow (2018) have also applied a novel psychometric approach to determining risk for amnestic and non-amnestic variants of mild cognitive impairment (MCI) in a large community-based sample, based on performance across a wide battery of neuropsychological tests ($N = 461$). In this study, based on performance across a large sample of neuropsychological tasks (see Payne & Stine-Morrow, 2014 for more information), participants were classified for increased risk for amnestic (memory-specific) or (single- or multi-domain) non-amnestic MCI. Adults displaying significant impairment (at least 1 SD below average) in at least one non-memorial cognitive domain (such as reasoning, or visuospatial ability) were classified as at an increased risk for non-amnestic MCI, while participants who showed impairment only in tasks assessing episodic memory (<1 SD below average) were classified as at increased risk for amnestic MCI. Such psychometric classification schemes have shown to be predict longitudinal conversion to AD (Cook et al., 2013), and show high overlap with clinical consensus (Clark et al., 2013).

In this study, Payne and Stine-Morrow observed an interesting dissociation between personality traits and risk for amnestic and non-amnestic variants of MCI. Those at higher risk for non-amnestic MCI displayed lower trait openness and also higher agreeableness. However, those with a memory-specific impairment had reliably lower conscientiousness. Interestingly, this was also one of the few studies to implicate trait agreeableness as a risk factor for pathological cognitive aging (see also Terracciano et al., 2014), and suggest that such multivariate assessments of cognitive impairment may have unique predictive capacity compared to typical correlational approaches to assess personality and specific cognitive domains. Overall, these findings suggest that personality traits may be useful indicators for different clinically-relevant risk factors of cognitive impairment in aging.

11.2.3 Memory Self-Efficacy in Aging

Although most of the existing work on personality-cognition relationships has focused on the Big Five personality traits, likely given their broad application across samples (John, Naumann, & Soto, 2008), there are a number of other important dispositional traits to consider in differential studies of cognitive aging. For instance, individuals vary considerably in their beliefs and self-appraisals of their own cognitive capabilities. In older adulthood, context-independent self-appraisals of one's own memory status are a stable trait-like dimension (Bandura, 1989; Berry & West, 1993). Importantly, while memory beliefs correlate with memory performance, they are not solely reflective of accurate meta-cognitive appraisals of ones' own memory status. A meta-analysis of over 100 studies by Beaudoin and Desrichard (2011) estimated a reliable but quite modest relationship between subjective memory beliefs and objective memory performance ($r = .15$) with substantial heterogeneity in effect sizes across studies, which shows the relevance of self-concepts in understanding cognitive functioning.

A major topic in the literature on cognitive self-efficacy beliefs revolves around the domain generality of such 'trait' or 'global' measures (see e.g., Hertzog and Pearman, 2013; Hertzog McGuire, Horhota, & Jopp, 2010). Although memory self-efficacy beliefs appear to be relatively stable longitudinally, some studies have shown that global measures of memory beliefs are associated with depression and the endorsement of negative personality traits (e.g., neuroticism). This has led some to argue that global memory belief measures show relationships with observed cognitive performance in part due to developmental changes in emotional and personality characteristics over the lifespan (Hertzog and Pearman, 2013; Pearman & Storandt, 2004).

Building on this literature, some studies (Amariglio, Townsend, Grodstein, Sperling, & Rentz, 2011; Haslam et al., 2012; Jopp & Hertzog, 2007; Payne et al., 2017) have begun exploring broader relationships between memory beliefs and cognitive performance, outside of the narrow scope of memory performance alone. For example, Payne et al. (2017) used bi-factor structural equation models to parse the shared and independent variance among cognitive factors representing episodic memory, psychomotor speed, executive reasoning, and general fluid cognition (g) and further, to explore the relationship between these cognitive factors and concurrent memory self-efficacy beliefs in two large cohort studies of older adults (Senior Odyssey, Stine-Morrow et al., 2014, $N = 462$; ACTIVE, Ball et al., 2002, $N = 2802$). First, they examined the simultaneous prediction of memory beliefs from psychomotor speed, reasoning, and memory to test whether memory self-efficacy beliefs were specific to objective memory performance. They found modest relationships between memory beliefs and episodic memory status in both samples. However, accounting for general fluid cognition in the model substantially attenuated the domain-specific relationships between memory beliefs and memory function. In fact, general fluid cognition was the strongest predictor of memory beliefs in both samples. Such findings suggest that negative self-referent beliefs about memory are not necessarily reflective of specific age-related declines in memory per se. Instead, dispositional memory beliefs appear to reflect individual differences in cognitive status more generally, suggesting that self-reports of memory status have broader predictive validity than was previously believed. Indeed, this interpretation is consistent with research showing that memory beliefs are predictive of adherence and responsiveness in cognitive interventions among older adults that do not explicitly target memory performance (e.g., Payne et al., 2012).

11.2.4 Summary

The findings presented in the sections above highlight that personality and dispositional traits have a reliable relationship with cognitive aging. Not only do individual differences in personality predict concurrent cognitive functioning and longitudinal change in cognitive functioning in older adulthood, but certain personality traits may hold additional predictive validity in separating expected cognitive change in aging from non-normative trajectories of change as observed in MCI and AD. Finally, we argue that we need to move beyond the Big Five personality traits often studied in younger adulthood, and to consider personality constructs that are more relevant to later adulthood, including memory self-efficacy. In the next section, we discuss how personality can serve as a precursor of both risk and protective factors in cognitive health over the lifespan.

11.3 Personality as a Precursor to Risk/Protective Factors in Cognitive Aging

In a review, Smith (2016) highlighted important candidate factors for offsetting cognitive decline in aging, including physical fitness, positive health behaviors (e.g., smoking cessation), intellectual engagement, diet, and social engagement. These important health behaviors serve as protective factors for maintaining cognitive health in older adulthood. Importantly, research on personality and health behaviors has consistently established personality traits that are predictive of these important behavioral factors (see Hill & Allemand, this edition; Mroczek & Weston, this edition). In short, personality is likely to play a critical role as a precursor of engagement across many of these domains (Hill & Payne, 2017). For example, higher conscientiousness predicts positive health behaviors, such as lower rates of drinking, smoking, and obesity. It may not be surprising then that conscientiousness has emerged as an all-cause predictor of mortality (Turiano, Chapman, Gruenewald, & Mroczek, 2015). Related to this, age-of-onset of several critical diseases can be predicted by individual differences in trait Conscientiousness, Neuroticism, and Openness (Weston, Hill, & Jackson, 2015).

Small but consistent effects of conscientiousness, openness, extraversion and neuroticism and physical exercise have been established (Wilson & Dishman, 2015). Such a relationship is critical for understanding cognitive and brain health given the strong associations between cardiovascular health, exercise, and neurological integrity in aging (Kramer and Hillman, 2006). Additionally, health literacy and medical adherence are in-and-of themselves complex cognitive behaviors that are both predicted by changes in cognitive functioning and predict sustained cognitive health throughout adulthood (Chin et al., 2015). Previous work (Hill & Roberts, 2011) has shown that higher levels of conscientiousness in older adulthood are associated with better medication adherence, again suggesting that personality traits may be important precursors to health behaviors to maintain effective cognitive functioning throughout adulthood.

Building on Hill and Payne (2017), we propose a conceptual model in which health-relevant personality traits reflect stable dispositional factors that contribute to the propensity to engage in positive health behaviors on a daily basis and maintain such behaviors over time in the face of daily stressors and life events. These positive health and lifestyle factors are likely to accumulate over decades, yielding a net preventive effect on cognitive decline across a range of domains. Although such a model has not been thoroughly empirically tested, there are findings from some domains, notably in intellectual activity engagement, that are consistent with this model (e.g., Jopp and Hertzog, 2007, Soubelet & Salthouse, 2011; Jackson et al., in review).

Findings from the Women's Health and Aging study (Carlson et al., 2012) suggested one such potential pathway between activity engagement and cognitive health. In this study, they showed that individuals who self-reported a greater *variety* of activities, rather than the frequency of engagement in specific intellectually stimulating or cognitively demanding activities, showed improved cognition. A greater variety of participation in activities, regardless of cognitive challenge, was associated with an 8–11% reduction in the risk of impairment in verbal memory and global cognitive outcomes. Furthermore, participation in a variety of lifestyle activities was more predictive than frequency or level of cognitive challenge for significant reductions in risk of incident impairment on measures sensitive to cognitive aging and risk for dementia.

Recently, Jackson and colleagues (under review) revisited the question of whether activity diversity is related to cognitive health outcomes in older adulthood and further examined whether increased activity diversity may be one putative mechanism underlying the relationship between trait openness and cognitive functioning in older adulthood. First, they replicated the finding from Carlson and colleagues showing that the total number of different activities one participates in is more strongly related to a number of cognitive outcomes compared to just the frequency of time spent in any set of activities. This even held true when examining a subset of activities that were judged as most cognitively engaging. Activity diversity reliably predicted composite measures of inductive reasoning, divergent thinking (i.e., ideational fluency), psychomotor speed, verbal ability, and episodic memory, with effect sizes ranging from .23 to .27. Importantly, they also found that trait openness was positively related to reported activity diversity, but not with the total number of hours spent in all activities or in cognitively demanding activities alone. Putting this altogether, they used bootstrapped mediation tests to examine whether activity diversity mediated the relationship between openness and each cognitive composite. They found evidence that activity diversity partially mediated the relationship between openness and three cognitive composites—inductive reasoning, processing speed, and episodic memory. These findings are valuable as first steps in establishing data consistent with the conceptual model discussed above and provide evidence in support of the idea that personality traits may serve as precursors to positive lifestyle factors that promote cognitive health in older adulthood. At the same time, considerable work remains to further examine such health-behavior and lifestyle mediational pathways between personality and cognition, and examine such relationships longitudinally to attempt to address causal mechanisms.

11.3.1 Summary

It is crucial for researchers to more thoughtfully consider the role of personality in future research on cognitive aging, not simply because certain personality traits are correlates of cognitive functioning in older adulthood but because personality may serve as an important and potentially actionable precursor to many health-related behaviors and cognitive health risks in adulthood. In the above section, we highlighted a theoretical model whereby certain personality traits may impact daily life behaviors that have net positive or negative effects on cognitive functioning in adulthood. Although only a small number of studies have empirically tested this model, we view it as a fruitful way forward to understanding the multiple pathways that link personality constructs to patterns of cognitive decline in aging, which will eventually help us to understand how we may personally individualize interventions to promote cognitive health in adulthood. This is the topic of our final section, below.

11.4 Personality as a Promoter of Cognitive Intervention

In the prior section, we considered the degree to which personality may shape patterns of positive health behaviors and activities over the lifespan in such a way to promote cognitive health. Another important consideration through which personality may relate to cognitive health is through promoting engagement in targeted interventions designed to improve the cognitive functioning of older adults. In fact, there is a small but growing literature highlighting the seemingly powerful role that individual differences in personality traits play in cognitive interventions. Below, we review several recent and diverse examples that suggest individual differences in personality could moderate the efficacy of interventions designed to target cognitive functioning in healthy adults of varying ages, as well as those that target cognitive symptomology in adults with dementia.

In the clinical literature, the influence of personality on activity participation among individuals with dementia has been consistently demonstrated (Kolanowski, Litaker, Buettner, Moeller, & Costa, 2011; Hill, Kolanowski, Fick, Chinchilli, & Jablonski, 2014). For example, Kolanowski et al. (2011) reported results from a randomized clinical trial in which 128 cognitively impaired nursing-home residents were randomly assigned to an individually-designed activity engagement intervention that was tailored to either their functional level, personality characteristics, or both personality and functional level together. Personality-based activities were defined based on self-reports of certain facets of extraversion (gregariousness, assertiveness, activity, excitement seeking) and openness (fantasy, aesthetics, feelings, ideas). For example, if a participant scored high on excitement seeking and gregariousness, they could be assigned to a novel tetherball game with other individuals. Patients randomly assigned to the personality-tailored activities reported greater engagement, alertness, and attention as well as reduced agitation and passivity compared to the other groups, suggesting that tailoring individualized activities based on personality may result in better adherence-related outcomes, which are critical for yielding long-term benefits.

Similar findings were reported by Hill et al. (2014) who examined the moderating effects of personality traits on cognitive function following an individualized activity engagement intervention among a sample of individuals with delirium superimposed on dementia. The activity engagement intervention was administered individually each day for 30 minutes for a maximum 30 days and entailed customized mentally challenging recreational activities that became incrementally difficult over the course of the intervention. They found that participants scoring high on agreeableness showed differentially improved delayed recall, and those with lower extraversion showed improved executive functioning following the intervention. Moreover, openness, agreeableness, and conscientiousness were associated with greater engagement in the activity intervention. Collectively, these findings suggest that personality traits are important to consider when selecting and tailoring cognitive interventions for older individuals with cognitive impairment.

In the literature on individualized cognitive training in healthy young and older adults, personality traits have been shown to be consistent predictors of individual differences in training adherence and the degree of improvement due to training. Double and Birney (2016) examined the influence of personality and self-referent beliefs on cognitive training adherence and performance outcomes in a large sample of 831 older Australians who self-selected into a commercial brain training program. They found that openness, need for cognition, and age predicted continuation in the computerized training program. Moreover, they observed that openness, implicit theories of intelligence, and age independently predicted task performance.

These findings suggest that one pathway through which personality can impact improvements in performance is through predicting program adherence (see also Payne et al., 2012).

Studer-Luethi, Jaeggi, Buschkuehl, and Perrig (2012) examined individual differences in trait conscientiousness and neuroticism as moderators of an intensive working memory training intervention among 112 young adults (mean age = 19.5). Participants were randomly assigned to either a single or dual *n-back* working memory training task or to a no-contact control group. The n-back task is a continuous performance task in which participants are instructed to monitor a series of stimuli (e.g., letters or digits) and respond if the current stimulus is identical to the one presented *n* trials previously (e.g., two-trials previously). In the dual n-back variant, participants have to simultaneously monitor two stimulus streams (e.g., visually presented digits and aurally presented letters). Participants in the challenging dual n-back task who scored high in neuroticism showed lower overall gains in both near transfer tasks (e.g., n-back) and far transfer tasks (e.g., fluid intelligence), as well as reporting lower training enjoyment overall. However, in the single n-back group, participants high in neuroticism showed *better* performance in the simpler WM training intervention. These findings suggest that the high demand of the dual n-back training task led subjects with high levels of neuroticism to perform sub-optimally, derailing potential transfer processes. A similar training x trait interaction was observed as a function of conscientiousness, where higher conscientiousness was associated with high immediate training improvement in the single n-back task, as well as greater improvement in near transfer measures, but at the same time, reduced far-transfer performance. The authors argued that these findings suggested that participants with high conscientiousness may have developed task-specific skills to perform as efficiently as possible, thus reducing any potential generalizing effects of the training. Thus, individual differences in personality had a substantial influence on the cognitive strategies adopted during training, leading to differential responsiveness to simple versus complex training paradigms.

Payne et al. (2012) examined the moderating effects of memory self-concept on cognitive training adherence and outcomes in a non-memorial domain, that of inductive reasoning. Although a number of researchers had previously examined memory training and self-efficacy beliefs (West, Bagwell, & Dark-Freudeman, 2008; McDougall, 2009), little work had previously examined the predictive utility of trait memory self-efficacy beliefs on interventions not explicitly targeting memory outcomes. In this study (Payne et al., 2012), 105 older-adult participants (mean age = 72.9) were randomly assigned to a no-contact control group or a home-based inductive reasoning intervention (Margrett & Willis, 2006), in which participants were trained in recognizing novel patterns and using these patterns to solve problems. Reasoning training materials were designed to increase in level of difficulty from week to week.

Overall, the training group showed substantial improvements in reasoning but no evidence for transfer to other cognitive domains, consistent with prior reasoning interventions (see e.g., Ball et al., 2002). Latent change score models were used to examine individual differences in latent reasoning change as a function of the intervention. Importantly, within the training group only, better memory self-belief was associated with improved reasoning outcomes, suggesting that self-referent memory beliefs are predictive of intervention responsiveness. Furthermore, those adults with more positive beliefs were shown to allocate more time to the training materials over the course of the intervention, whereas adults with lower beliefs began allocating less time to the training over the course of the intervention. Such findings indicate that self-referential beliefs about cognitive potential may be an important factor contributing to intervention responsiveness in adulthood. Importantly, such findings may be more pronounced among interventions that require greater self-regulation and allocation of time and effort, as some more stereotyped lab-based training interventions (e.g., useful field of view training for processing speed) do not show such heterogeneity in training outcomes (Sharpe, Holup, Hansen, & Edwards, 2014).

Finally, we discuss findings from the Senior Odyssey Intervention (Stine-Morrow et al., 2008, 2014), a socially-enriched activity engagement intervention designed to improve cognitive health. This study serves as one of the largest and best tests of the role of personality characteristics as moderators of treatment effects. In the Senior Odyssey intervention, older adults are embedded in a cognitively and socially stimulating environment to promote activity engagement. Participants work in collaboration with the Odyssey of the Mind (OOTM) Program, which has been in existence since the late 1970's for university-level students. Participants compete in team-based collaborative and creative problem-solving activities over an entire 'season' of the OOTM program. An example of a long-term problem is, designing a play or designing structures out of balsa wood that can hold the most weight possible. Such problems require long-term planning, testing, and revision in a socially engaging environment to achieve desired solutions. Two phases of the study were conducted. In the first preliminary study (Stine-Morrow et al., 2008), 181 highly active adults were randomly assigned to the intervention or to a no-contact control group. This group showed broad improvement across a number of cognitive domains, but not substantial evidence of individual differences in training responsiveness.

In the second phase of the study (Stine-Morrow et al., 2014), over 400 'low activity' older adults were targeted (less than 10 hours of scheduled activities a week), as this training may be differentially beneficial for less engaged adults. Participants were randomly assigned to the OOTM program, or to either a waitlist control group or an active control group that completed home-based reasoning training (Willis et al., 2006; Payne et al., 2012). The lower-activity sample showed very different responsiveness to the training, showing effects that resembled specific training benefits in ideational fluency (a key component of the OOTM problems). However, participants also showed substantial individual differences in responsiveness to the intervention.

Moderators of individual differences in fluency change as a function of the intervention were examined. Here, effects of age and initial cognitive status predicted training improvement, but importantly, independent effects of trait openness and social network size also moderated the magnitude of improvement in the intervention group, consistent with the idea that these features are critical to engage in this kind of broad and open-ended social problem-solving type activity. That is, in socially-mediated real-world activities, social and personality factors may play a larger role in engagement and adherence than in other types of more contrived or individualized interventions. Notably, the effect sizes for openness and social network size were the same magnitude as the moderating influence of baseline age and initial cognitive status, showing again the power of personality and dispositional individual differences characteristics in predicting cognitive intervention outcomes in aging.

11.4.1 Summary

The above findings strongly suggest that personality traits and dispositional beliefs in adulthood can have a substantial impact on responsiveness to cognitive training interventions. Individual differences in traits at the beginning of the interventions appear to influence a number of factors, including the strategies adopted to complete the tasks, belief and enjoyment in the intervention, and maintained adherence and engagement with the tasks, and degree of cognitive change as a result of the intervention. Some work suggests that individual differences may play a larger role in cases where the intervention is open-ended and self-guided, requiring more self-regulation, motivation, and social engagement. For example, in lab-based computerized cognitive training programs individual differences in personality traits may only moderately impact training outcomes, as factors like conscientiousness and cognitive self-efficacy may moderate how participants maintain adherence to the training. However, in less structured interventions, such as home-based or socially-interactive group-based interventions, such as Senior Odyssey (Stine-Morrow et al., 2008, 2014), which require additional self-motivation, planning, and engagement in open-ended problem solving, individual personality characteristics appear to play a larger role in determining training responsiveness.

11.5 Future Directions

The goal of this chapter was to systematically review the multiple pathways through which personality traits and cognitive functioning interact in aging and adult development. Importantly, the above review shows a small but rapidly growing literature highlighting the importance of personality traits in shaping trajectories of normal cognitive functioning and cognitive health maintenance in old age. As such, we argue that cognitive aging research should more seriously integrate personality science into basic research on cognitive change in normal and pathological aging as well as applied research on interventions to promote cognitive health in adulthood. Below, we briefly outline several ways in which personality traits can be readily integrated into adult developmental research on cognitive and brain health.

Ongoing longitudinal studies of cognitive and brain health should integrate personality theory and assessment into their existing protocols. Aside from simply including a brief Big Five measure into existing measurement batteries, as is currently the convention, such studies should explore the predictive validity of including behavioral and observer-report assessments into existing and planned future studies, as well as examining facet-level differences, and more age-relevant traits, such as self-referent cognitive beliefs. Likewise, intervention and training studies should consider the measurement of personality not only at the onset of assessment (e.g., as a moderator of cognitive change, e.g., Payne et al., 2012), but also longitudinally, to examine whether and to what degree such interventions may change personality and associated behaviors (e.g., Jackson, Hill, Payne, Roberts, & Stine-Morrow, 2012). Such intervention-related personality change, if observed, may result in more long-term effectiveness than the direct cognitive benefits of the intervention. Indeed, the effects of short-term cognitive interventions typically do not persist far beyond the conclusion of the intervention (see Rebok et al., 2014). However, if certain cognitive interventions can lead to changes in factors like openness, and conscientiousness, this personality trait change may result in longer-term and stable changes to behaviors, such as seeking out novel and intellectually engaging experiences, which may lead to broader and more lasting change in cognition than the direct but more transient cognitive benefits.

Moreover, personality science can be more thoroughly integrated in person-centered approaches to promote cognitive resilience in aging. As reviewed above, some promising examples of personality-tailored interventions have been applied in clinical interventions for dementia. However, very little work has attempted to integrate personality and dispositional belief change directly into cognitive interventions. One exception to this comes from West et al. (2008), who have implemented memory training interventions that are supplemented with tasks to improve perceived memory self-efficacy. Not only can such multifaceted approaches to intervening along multiple fronts prove successful (e.g., see Aschwanden & Allemand's theoretical model in this edition), but future interventions should additionally consider individually adjusting interventions to individual belief and trait profiles. For example, given the work reviewed above, namely that older adults with low self-efficacy and high neuroticism do not respond as strongly to cognitive training interventions, it is worth considering how such baseline information could be used to supplement such interventions, for example by increasing task engagement or 'gamification' or providing longer training intervals. Given that a one-size-fits-all approach is likely ineffective for cognitive interventions, such future work should consider such personality-integrated individually-adaptive interventions.

With the considerable public discussion over the last several decades on the 'graying of America', the conversation has focused largely on tools needed to stave off cognitive decline and improve brain health. Importantly, this view of cognitive and brain health largely leaves behind attention to *the person*. Consideration of the substantial heterogeneity in personality, beliefs, and dispositions is critical for a complete understanding of optimal adult life span development. We propose a shift in cognitive aging research, whereby we not only take into account normative and non-normative age-related changes (and variability in changes), in cognitive functioning and personality, but consider the dynamic interplay between these factors that have historically been examined in isolation. Such integration will prove valuable not only for forwarding basic research on personality and cognitive development, but will provide novel inroads for non-traditional approaches to improve cognitive health and maintain cognitive resilience over the adult lifespan.

References

Ackerman, P. L., & Heggestad, E. D. (1997). Intelligence, personality, and interests: Evidence for overlapping traits. *Psychological Bulletin, 121*(2), 219.

Amariglio, R. E., Townsend, M. K., Grodstein, F., Sperling, R. A., & Rentz, D. M. (2011). Specific subjective memory complaints in older persons may indicate poor cognitive function. *Journal of the American Geriatrics Society, 59*(9), 1612–1617.

Ball, K., Berch, D. B., Helmers, K. F., Jobe, J. B., Leveck, M. D., Marsiske, M., … Unverzagt, F. W. (2002). Effects of cognitive training interventions with older adults: A randomized controlled trial. *Journal of the American Medical Association, 288*(18), 2271–2281.

Bandura, A. (1989). Regulation of cognitive processes through perceived self-efficacy. *Developmental Psychology, 25*(5), 729–735.

Beaudoin, M., & Desrichard, O. (2011). Are memory self-efficacy and memory performance related? A meta-analysis. *Psychological Bulletin, 137*(2), 211.

Berry, J. M., & West, R. L. (1993). Cognitive self-efficacy in relation to personal mastery and goal setting across the life span. *International Journal of Behavioral Development, 16*(2), 351–379.

Bopp, K. L., & Verhaeghen, P. (2005). Aging and verbal memory span: A meta-analysis. *The Journals of Gerontology Series B: Psychological Sciences and Social Sciences, 60*(5), P223–P233.

Braver, T. S., & West, R. (2008). Working memory, executive control, and aging. *The Handbook of Aging and Cognition, 3*, 311–372.

Carlson, M. C., Parisi, J. M., Xia, J., Xue, Q. L., Rebok, G. W., Bandeen-Roche, K., & Fried, L. P. (2012). Lifestyle activities and memory: Variety may be the spice of life. The women's health and aging study II. *Journal of the International Neuropsychological Society, 18*(2), 286–294.

Chamorro-Premuzic, T., Furnham, A., & Moutafi, J. (2004). The relationship between estimated and psychometric personality and intelligence scores. *Journal of Research in Personality, 38*(5), 505–513.

Chin, J., Payne, B., Gao, X., Conner-Garcia, T., Graumlich, J. F., Murray, M. D., … Stine-Morrow, E. A. (2015). Memory and comprehension for health information among older adults: Distinguishing the effects of domain-general and domain-specific knowledge. *Memory, 23*(4), 577–589.

Clark, L. R., Delano-Wood, L., Libon, D. J., McDonald, C. R., Nation, D. A., Bangen, K. J., … Bondi, M. W. (2013). Are empirically-derived subtypes of mild cognitive impairment consistent with conventional subtypes? *Journal of the International Neuropsychological Society, 19*, 1–11. https://doi.org/10.1017/S1355617713000313

Cook, S. E., Marsiske, M., Thomas, K. R., Unverzagt, F. W., Wadley, V. G., Langbaum, J. B., & Crowe, M. (2013). Identification of mild cognitive impairment in ACTIVE: Algorithmic classification and stability. *Journal of the International Neuropsychological Society, 19*, 73. https://doi.org/10.1017/S1355617712000938

Curtis, R. G., Windsor, T. D., & Soubelet, A. (2015). The relationship between Big-5 personality traits and cognitive ability in older adults - a review. *Neuropsychology, Development, and Cognition. Section B, Aging, Neuropsychology and Cognition, 22*(1), 42–71.

Deary, I. J., Penke, L., & Johnson, W. (2010). The neuroscience of human intelligence differences. *Nature Reviews Neuroscience, 11*(3), 201.

Double, K. S., & Birney, D. P. (2016). The effects of personality and metacognitive beliefs on cognitive training adherence and performance. *Personality and Individual Differences, 102*, 7–12.

Eckert, M. A. (2011). Slowing down: Age-related neurobiological predictors of processing speed. *Frontiers in Neuroscience, 5*, 1–13.

Federmeier, K. D., Van Petten, C., Schwartz, T. J., & Kutas, M. (2003). Sounds, words, sentences: Age-related changes across levels of language processing. *Psychology and Aging, 18*(4), 858.

Haslam, C., Morton, T. A., Haslam, S. A., Varnes, L., Graham, R., & Gamaz, L. (2012). "When the age is in, the wit is out": Age-related self-categorization and deficit expectations reduce performance on clinical tests used in dementia assessment. *Psychology and Aging, 27*(3), 778.

Hertzog, C., Kramer, A. F., Wilson, R. S., & Lindenberger, U. (2008). Enrichment effects on adult cognitive development: Can the functional capacity of older adults be preserved and enhanced? *Psychological Science in the Public Interest, 9*(1), 1–65.

Hertzog, C., McGuire, C. L., Horhota, M., & Jopp, D. (2010). Does believing in "use it or lose it" relate to self-rated memory control, strategy use, and recall? *The International Journal of Aging and Human Development, 70*(1), 61–87.

Hertzog, C., & Pearman, A. (2013). Memory complaints in adulthood and old age. In *The SAGE Handbook of Applied Memory* (pp. 423–443). London, UK: SAGE.

Hill, N. L., Kolanowski, A. M., Fick, D., Chinchilli, V. M., & Jablonski, R. A. (2014). Personality as a moderator of cognitive stimulation in older adults at high risk for cognitive decline. *Research in Gerontological Nursing, 7*(4), 159–170.

Hill, P. L. & Payne, B. R. (2017). *Don't forget the person when promoting healthy cognitive aging: Comment on Smith (2016)*. American psychologist, *72*(4), 390–392

Hill, P. L., & Roberts, B. W. (2011). The role of adherence in the relationship between conscientiousness and perceived health. *Health Psychology, 30*(6), 797.

Jackson, J. J., Hill, P. L., Payne, B. R., Roberts, B. W., & Stine-Morrow, E. A. (2012). Can an old dog learn (and want to experience) new tricks? Cognitive training increases openness to experience in older adults. *Psychology and Aging, 27*(2), 286.

John, O. P., Naumann, L. P., & Soto, C. J. (2008). Paradigm shift to the integrative big five trait taxonomy. *Handbook of Personality: Theory and Research, 3*(2), 114–158.

Jopp, D., & Hertzog, C. (2007). Activities, self-referent memory beliefs, and cognitive performance: Evidence for direct and mediated relations. *Psychology and Aging, 22*(4), 811.

Kolanowski, A., Litaker, M., Buettner, L., Moeller, J., & Costa, P. T., Jr. (2011). A randomized clinical trial of theory-based activities for the behavioral symptoms of dementia in nursing home residents. *Journal of the American Geriatrics Society, 59*(6), 1032–1041.

Kramer, A. F., & Hillman, C. H. (2006). Aging, physical activity, and neurocognitive function. *Psychobiology of Physical Activity, 45*, 59.

Lorge, I. (1940). Intelligence: Its nature and nurture. *The 39th Yearbook, 39*(Pt. I), 275–281.

Lien, M. C., Allen, P. A., Ruthruff, E., Grabbe, J., McCann, R. S., & Remington, R. W. (2006). Visual word recognition without central attention: evidence for greater automaticity with advancing age. *Psychology and Aging, 21*(3), 431.

Luchetti, M., Terracciano, A., Stephan, Y., & Sutin, A. R. (2016). Personality and cognitive decline in older adults: Data from a longitudinal sample and meta-analysis. *The Journals of Gerontology Series B: Psychological Sciences and Social Sciences, 71*(4), 591–601.

Manly, J. J., Touradji, P., Tang, M. X., & Stern, Y. (2003). Literacy and memory decline among ethnically diverse elders. *Journal of Clinical and Experimental Neuropsychology, 25*(5), 680–690.

Margrett, J. A., & Willis, S. L. (2006). In-home cognitive training with older married couples: Individual versus collaborative learning. *Aging, Neuropsychology, and Cognition, 13*(2), 173–195.

McDougall, G. J., Jr. (2009). A framework for cognitive interventions targeting everyday memory performance and memory self-efficacy. *Family & Community Health, 32*(1 Suppl), S15.

Moutafi, J., Furnham, A., & Crump, J. (2003). Demographic and personality predictors of intelligence: A study using the NEO personality inventory and the Myers–Briggs type indicator. *European Journal of Personality, 17*(1), 79–94.
Moutafi, J., Furnham, A., & Paltiel, L. (2004). Why is conscientiousness negatively correlated with intelligence? *Personality and Individual Differences, 37*(5), 1013–1022.
Murray, A. L., Johnson, W., McGue, M., & Iacono, W. G. (2014). How are conscientiousness and cognitive ability related to one another? A re-examination of the intelligence compensation hypothesis. *Personality and Individual Differences, 70*, 17–22.
Park, D. C., Smith, A. D., Lautenschlager, G., Earles, J. L., Frieske, D., Zwahr, M., & Gaines, C. L. (1996). Mediators of long-term memory performance across the life span. *Psychology and Aging, 11*(4), 621.
Payne, B. R., Gao, X., Noh, S. R., Anderson, C. J., & Stine-Morrow, E. A. (2012). The effects of print exposure on sentence processing and memory in older adults: Evidence for efficiency and reserve. *Aging, Neuropsychology, and Cognition, 19*(1–2), 122–149.
Payne, B. R., Gross, A. L., Hill, P. L., Parisi, J. M., Rebok, G. W., & Stine-Morrow, E. A. (2017). Decomposing the relationship between cognitive functioning and self-referent memory beliefs in older adulthood: what's memory got to do with it? *Aging, Neuropsychology, and Cognition, 24*(4), 345–362.
Payne, B. R., & Stine-Morrow, E. A. (2014). Risk for mild cognitive impairment is associated with semantic integration deficits in sentence processing and memory. *Journals of Gerontology Series B: Psychological Sciences and Social Sciences, 71*(2), 243–253.
Payne, B. R., & Stine-Morrow, E. A. L. (2018). Personality traits and markers of cognitive impairment in aging. *Innovation in Aging, 2*(Suppl 1), 749.
Pearman, A., & Storandt, M. (2004). Predictors of subjective memory in older adults. *The Journals of Gerontology Series B: Psychological Sciences and Social Sciences, 59*(1), P4–P6.
Rammstedt, B., Danner, D., & Martin, S. (2016). The association between personality and cognitive ability: Going beyond simple effects. *Journal of Research in Personality, 62*, 39–44.
Rebok, G. W., Ball, K., Guey, L. T., Jones, R. N., Kim, H. Y., King, J. W., … Willis, S. L. (2014). Ten-year effects of the advanced cognitive training for independent and vital elderly cognitive training trial on cognition and everyday functioning in older adults. *Journal of the American Geriatrics Society, 62*(1), 16–24.
Salthouse, T. A. (1996). The processing-speed theory of adult age differences in cognition. *Psychological Review, 103*, 403–428.
Salthouse, T. A. (2019). Trajectories of normal cognitive aging. *Psychology and Aging, 34*(1), 17.
Salthouse, T. A., & Ferrer-Caja, E. (2003). What needs to be explained to account for age-related effects on multiple cognitive variables? *Psychology and Aging, 18*(1), 91.
Salthouse, T. A., & Madden, D. J. (2013). Information processing speed and aging. In *Information processing speed in clinical populations* (Vol. 221). London, UK: Taylor and Francis.
Schaie, K. W. (1996). Intellectual development in adulthood. *Handbook of the Psychology of Aging, 4*, 266–286.
Schaie, K. W., Willis, S. L., & Caskie, G. I. (2004). The Seattle longitudinal study: Relationship between personality and cognition. *Aging Neuropsychology and Cognition, 11*(2–3), 304–324.
Segerstrom, S. C. (2018). Personality and incident Alzheimer's disease: Theory, evidence, and future directions. *The Journals of Gerontology: Series B, 73*(1), 1–184.
Sharpe, C., Holup, A. A., Hansen, K. E., & Edwards, J. D. (2014). Does self-efficacy affect responsiveness to cognitive speed of processing training? *Journal of Aging and Health, 26*(5), 786–806.
Smith, G. E. (2016). Healthy cognitive aging and dementia prevention. *American Psychologist, 71*(4), 268.
Soubelet, A., & Salthouse, T. A. (2011). Personality–cognition relations across adulthood. *Developmental Psychology, 47*(2), 303.
Stern, Y. (2012). Cognitive reserve in ageing and Alzheimer's disease. *The Lancet Neurology, 11*(11), 1006–1012.

Stine-Morrow, E. A., Parisi, J. M., Morrow, D. G., & Park, D. C. (2008). The effects of an engaged lifestyle on cognitive vitality: A field experiment. *Psychology and Aging, 23*(4), 778.

Stine-Morrow, E. A., Payne, B. R., Roberts, B. W., Kramer, A. F., Morrow, D. G., Payne, L., … Janke, M. C. (2014). Training versus engagement as paths to cognitive enrichment with aging. *Psychology and Aging, 29*(4), 891.

Studer-Luethi, B., Jaeggi, S. M., Buschkuehl, M., & Perrig, W. J. (2012). Influence of neuroticism and conscientiousness on working memory training outcome. *Personality and Individual Differences, 53*(1), 44–49.

Sutin, A. R., Stephan, Y., Damian, R. I., Luchetti, M., Strickhouser, J. E., & Terracciano, A. (2019). Five-factor model personality traits and verbal fluency in 10 cohorts. *Psychology and Aging, 34*(3), 362.

Sutin, A. R., Terracciano, A., Kitner-Triolo, M. H., Uda, M., Schlessinger, D., & Zonderman, A. B. (2011). Personality traits prospectively predict verbal fluency in a lifespan sample. *Psychology and Aging, 26*(4), 994.

Terracciano, A., & Sutin, A. R. (2019). Personality and Alzheimer's disease: An integrative review. *Personality Disorders: Theory, Research, and Treatment, 10*(1), 4–12.

Terracciano, A., Sutin, A. R., An, Y., O'Brien, R. J., Ferrucci, L., Zonderman, A. B., & Resnick, S. M. (2014). Personality and risk of Alzheimer's disease: New data and meta-analysis. *Alzheimer's & Dementia, 10*, 179–186.

Turiano, N. A., Chapman, B. P., Gruenewald, T. L., & Mroczek, D. K. (2015). Personality and the leading behavioral contributors of mortality. *Health Psychology, 34*(1), 51.

Vaportzis, E., & Gow, A. J. (2018). People's beliefs and expectations about how cognitive skills change with age: Evidence from a UK-wide aging survey. *The American Journal of Geriatric Psychiatry, 27*(10), 1035–1160.

Verhaeghen, P. (2003). Aging and vocabulary score: A meta-analysis. *Psychology and Aging, 18*(2), 332.

West, R. L., Bagwell, D. K., & Dark-Freudeman, A. (2008). Self-efficacy and memory aging: The impact of a memory intervention based on self-efficacy. *Neuropsychology, Development, and Cognition Section B, Aging, Neuropsychology and Cognition, 15*, 302–329.

Weston, S. J., Hill, P. L., & Jackson, J. J. (2015). Personality traits predict the onset of disease. *Social Psychological and Personality Science, 6*(3), 309–317.

Willis, S. L., Tennstedt, S. L., Marsiske, M., Ball, K., Elias, J., Koepke, K. M., … Wright, E. (2006). Long-term effects of cognitive training on everyday functional outcomes in older adults. *JAMA, 296*(23), 2805–2814.

Wilson, K. E., & Dishman, R. K. (2015). Personality and physical activity: A systematic review and meta-analysis. *Personality and Individual Differences, 72*, 230–242.

World Health Organization. (2015). *World report on ageing and health*. Geneva, Switzerland: World Health Organization.

Chapter 12
A Lifespan Perspective on the Interconnections Between Personality, Health, and Optimal Aging

Daniel K. Mroczek, Sara J. Weston, and Emily C. Willroth

12.1 Introduction

There is renewed interest in the concept of Healthy Aging, which had known previous incarnations with such labels as Successful Aging (Baltes & Baltes, 1990; Rowe & Kahn, 1997) and Optimal Aging (Aldwin, Spiro, & Park, 2006). Whether it is called healthy, successful, or optimal, the concept refers to aging that is not pathological (e.g., early aging) and at the same time is better than normal aging. We invoke images of a physically active 90 year old or a cognitively intact 80 year old to illustrate successful or healthy, as opposed to normal or even pathological aging. Perhaps the key question in the study of healthy aging is how to promote it or bring it about to some extent. Of course, genetics and social structural variables (e.g., family SES, living in a rich country in the developed world, living in an environment that facilitates healthy habits versus one that promotes obesity) influence the ability of individuals to age in a healthful manner. Yet other individual factors such as personality traits also play a role in whether a person is a successful ager. The seeds of Healthy Aging are sown early in life, through the health choices we make in younger adulthood and midlife. Cumulative effects of good health behaviors throughout one's life often pay off in older adulthood, in the form of healthy aging, whereas poorer choices earlier in life can lead to suboptimal or worse, pathological aging (Mokdad, Marks, Stroup, & Gerberding, 2004; Peel, McClure, & Bartlett, 2005). Personality shapes those choices (Friedman, 2000; Smith, 2006; Turiano, Chapman, Gruenewald, & Mroczek, 2015).

D. K. Mroczek (✉) · E. C. Willroth
Northwestern University, Chicago, IL, USA
e-mail: daniel.mroczek@northwestern.edu

S. J. Weston
Department of Psychology, University of Oregon, Eugene, OR, USA

P. L. Hill, M. Allemand (eds.), *Personality and Healthy Aging in Adulthood*, International Perspectives on Aging 26,
https://doi.org/10.1007/978-3-030-32053-9_12

This chapter focuses on personality characteristics and healthy aging. First, we define healthy aging. Next, we suggest that personality traits throughout the lifespan play a central role in setting the stage for healthy aging. We discuss long-term explanatory chains that explicate how personality can have its effect on healthy aging not only in later life but also in earlier periods within the lifespan including childhood. Then, we describe more complex bidirectional and multidirectional associations between personality and health across the lifespan. Finally, we posit that understanding the complex associations between personality and health can help create more effective interventions to promote healthy aging that are targeted and tailored based upon people's personalities.

12.2 The Phenomenon of Healthy Aging

There are multiple definitions of healthy aging. These definitions typically include functional status, or the extent to which a person can function independently, take care of one's self, and contribute to society (World Health Organization, 2017). The World Health Organization (WHO) definition of functional status incorporates intrinsic capacity, which recognizes the multitude of mental and physical capacities that one can draw upon, as well as environments that support one's functional ability. In addition to functional status, many definitions of healthy aging also encompass the capacity to thrive and flourish in older adulthood (Baltes & Baltes, 1990). For example, remaining a productive older worker or travelling the world in later life would constitute successful or healthy aging. These definitions illustrate that healthy aging is a multi-faceted phenomenon that encompasses the holistic health and well-being of older adults.

12.2.1 David Barker and Long-Term Explanatory Chains: From Cradle to Grave

Personality characteristics impact many aspects of healthy aging in both distal and proximal ways. The causes can be found in factors that are nearer (in time) to older age and in factors that predate older adulthood by many decades. This idea rests upon the central tenets of lifespan development theory which maintains that developmental variables (like personality, physical functioning, cognitive functioning, emotions, etc.) are interconnected over long periods of time, even across the full life course (Baltes, 1987). The Barker fetal origins hypothesis that we discuss below is such an example. Other lifespan development principles are also highly relevant for the topic of personality and healthy aging in this chapter. For example, the tenet of multidirectionality in development (Baltes, 1987) bears directly upon the bidirectionality and feedback loop ideas we discuss later. The concept of development encompassing both gains and losses is quite relevant for the idea that the effects of a different personality dimension on health may sometimes comprise a gain (a net positive for health) but at other times frame a loss. The lifespan developmental concept of plasticity also frames the issue of personality and health. Personality characteristics and health variables are each plastic, or malleable across the lifespan. Further, when one changes, how does that effect the other? Does the association between personality and health become stronger under such circumstances, or does it become weaker? The association between the two variables is itself defined by the lifespan developmental principle of plasticity.

There are other tenets of lifespan developmental theory (e.g., historical embeddedness, contextualism), but multidirectionality, plasticity, and development as including both gains and losses, are the ones most relevant to how we conceptualize healthy aging in this chapter. A final tenet from Baltes (1987) is ontogenetic development, which we described earlier. It simply means development is a lifelong processes and events early in life are often interconnected in important ways to events that occur in later life. Sometimes this is framed in terms of proximal vs. distal associations. In this section, we describe these proximal and distal processes using a well-known example.

Distal associations can be defined by long-term explanatory chains that weave throughout the lifespan. The best example of a distal chain is the Fetal Origins of Late-Life Heart Disease, informally known as the Barker Hypothesis (Barker, 2004). It is one of the most well-known and well-documented of all long-term explanatory chains that describe a human lifespan biological or psychological process. The Barker Hypothesis states that very early life events, especially conditions in the womb, set the stage for late-life disease. It is astonishing to think that events that occur before birth, when we were in the fetal stage, could influence disease events that are as distal as heart disease in older adulthood. Decades separate the predictor and the outcome. Yet, the current thought on this topic posits the existence of many mediators or intervening variables that come between pre-birth womb conditions and heart failure that happens perhaps 5 or 6 decades post-birth. The hypothesis states that slower fetal growth rates lead to a "thrifty phenotype." In the short-term, this leads to lower birth weight, but also a lifelong slowdown of metabolic activity. This thrifty phenotype, with its metabolic limitations, does not necessarily lead to heart disease as long as a healthy, low-lipid diet is maintained. This thrifty phenotype is much less able to cope with the modern unhealthy diet, rich in fats and sugars, though. Such a diet, which is common in Western countries, leads to greater cardiometabolic problems in thrifty phenotype individuals.

We can visualize how a lifespan chain might operate with respect to the thrifty phenotype. It has its origins in the environment of the womb, perhaps due to nutrient deprivation of the mother due to poverty, wartime circumstances, or economic depression. It then manifests itself in low birth weight at the time of birth, and continues to manifest via slower metabolic processes in childhood, which if paired with a lipid-heavy diet would have obesiogenic outcomes (after the 1930s and 40s, caloric intake improved in many countries but perhaps not the quality of those calories). This in turn would lead to cardiometabolic problems in adulthood, eventually terminating in heart disease and earlier mortality in midlife and older adulthood. The Barker fetal origins model is a truly lifespan model of health, acknowledging that events from early life can influence healthy aging or other health outcomes that are years or decades distant.

The Barker Hypothesis is also a nice example of a lifespan chain model in which variables that are present early in life influence health much later in life through a multi-step chain that unfolds across the lifespan. It is also an example of Baltes' lifespan principle of ontogenetic development, or development as a cradle-to-grave, lifelong process. Abstracting this idea, we can conceive of how levels of a variable at the outset lead to an event or change in some other variable at a second time point, which in turn lead to an event or variable change at a third point, and on and on, until some endpoint (often much later in the lifespan) is reached. As an example, see Fig. 12.1: lower conscientiousness in young adulthood leads to the formation of sedentary habits, which leads to lack of physical fitness, which leads to further sedentary behavior, which leads to muscle atrophy and weight gain, which leads to joint problems, which leads to lack of mobility. Mobility issues are a central part of functional impairment, a key component of the WHO definition of non-healthy aging. The effects of personality may be very indirect, as this example illustrates. Here, personality initiates a very long-term developmental chain that unfolds over many decades, leading to an outcome relevant for healthy aging.

Yet there are examples of personality operating in a less distal, more proximal manner. For example, lower conscientiousness in late midlife (age 55–65) leads to many missed doctor appointments where the issue of sedentary lifestyle may have been addressed, leading to worse functional ability in late life. In other instances, proximal links between personality and health may be a single step in a longer

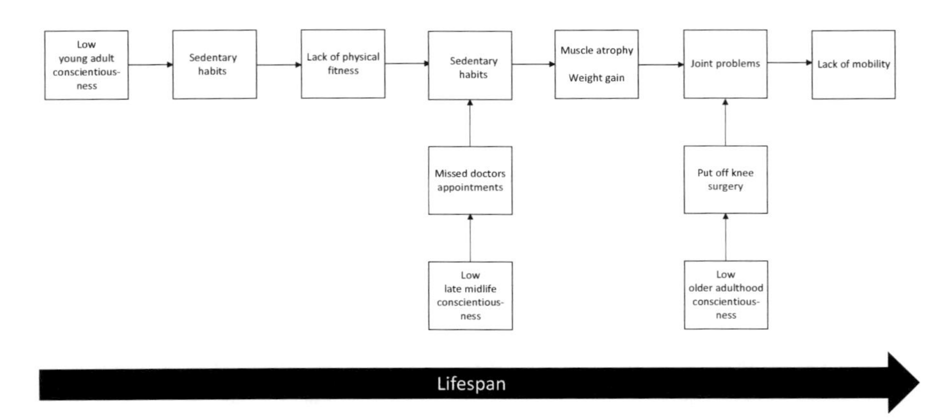

Fig. 12.1 An illustrative example of the varying effects of trait conscientiousness on health throughout the lifespan. Conscientiousness in young adulthood influences a trajectory of health behaviors; in this case, low conscientiousness leads to sedentary habits, which in turn influence a chain of worsening health and worsening health habits. Adulthood conscientiousness further influences this cycle, through missed doctor appoints and avoidance of procedures which may alter the health trajectory

chain. Weight gain in early and mid-adulthood puts strain on one's joints, leading to faster deterioration of cartilage in one's knees that requires knee replacement surgery. Low conscientiousness in older adulthood leads to putting off the knee surgery, increasing the number of years with lower functional ability, as well as worse surgical complications when the surgery is finally done (if it is undertaken at all). In this example, conscientiousness had a proximal effect on functional ability, but the chain of events leading to lower functional ability began with weight gain in early adulthood. As these examples illustrate, personality is involved in both distal and proximal processes which lead to heathy or unhealthy aging outcomes (Friedman, Kern, Hampson, & Duckworth, 2014; Hampson, Edmonds, Goldberg, Dubanoski, & Hillier, 2015; Shanahan, Hill, Roberts, Eccles, & Friedman, 2014).

12.3 Feedback Loops and Bidirectionality

Having described these hypothetical linear chains of causality, it is worth noting that more complex non-linear chains may be even more common. Undoubtedly, some causal chains may run neatly from A to B to C to D. This may hold when the elements of the chain (A, B, C, etc.) are comprised at least partially of events that have some kind of discrete ending. For example, a childhood disease (a discrete event) leads to a cognitive deficit (not an event but more akin to a trait or dimension) which in turn leads to less completed schooling and lower SES, which leads to poorer health, and ultimately to a heart attack (a discrete event). Yet other chains may not contain discrete events. Often, the elements in a chain are not events but rather are trait-like dimensions or perhaps chronic conditions that continue to exist even after they have exerted some effect on the next element in the chain. In such a circumstance, the leading element in the chain (A) exerts an effect on the second element (B), but A is still there even after it influences B. In turn, B may be a trait-like element itself that can exert an effect back on A, in addition to any effects B has on downstream elements (C, D, E). In such a feedback loop, B, which was changed by A, exerts an effect on A changing it further, which in turn feeds back and has an addition effect on B.

This undoubtedly sounds confusing. Yet if we think in a longitudinal and lifespan manner, it becomes much clearer. Think of the various elements in terms of time points or time segments. At time one, A exerts an effect on B, thereby changing B as it moves forward into time two. At time two, A and B are both extant, albeit B is changed from the previous time point. At time two, B exerts an effect on A. Now it is A's turn to change as it moves forward into time three. Of course, B is also extant at time three and is changed by A (which had been changed by B at time two, which had been changed by A at time one). This process of two or more variables influencing each other in a cyclical way as they move forward through time is often described by the term bidirectionality. In such cases, two or more variables influence each other in a cyclical way as they move forward through time.

An example of bidirectionality that has relevance for healthy aging research is the well-known association between wealth and health (Pollack et al., 2007). Wealth and health feed off each other and promote one another in a bidirectional manner, reaching a desirable terminus in late adulthood in the form of healthy aging and morbidity compression (i.e., a shorter period of poor health at the end of the lifespan). Higher earnings and greater net worth tend to lead to better health through various mediators, such as better health education and more skills to navigate the often byzantine world of health care and health insurance. In youth, when health issues tend to be at a minimum, higher family-of-origin SES often leads to higher levels of education. More education tends to be associated with greater knowledge about what causes health problems and what kinds of health behaviors can promote a healthier life. Education also tends to boost skill levels with respect to solving complex problems and processing complicated information. So greater wealth, as operationalized by higher family SES, can have a positive effect on health, through various mediators. Additionally, better health can feedback and boost one's wealth. One needs to be healthy to do complex and high-paying jobs, and to work long hours. Thus, the better health that was brought about by greater wealth in turn creates even greater wealth. Eventually, the twin action of wealth and health over a lifespan promote successful or healthy aging.

12.3.1 Bidirectionality in Personality and Healthy Aging

Turning to the bidirectional phenomenon in the context of personality and healthy aging, a handful of studies have demonstrated transactional or bidirectional relationships between personality traits and many other kinds of variables. Ironically, many of these other variables are regularly considered outcomes or endpoints. Yet in bidirectionality studies they are often shown not to be endpoints at all but rather part of a back-and-forth flux with a personality dimension that can go on for years, and presumably over the course of an entire lifespan.

Jackson (2011), for example, discovered that conscientiousness and academic engagement reinforced each other in a bidirectional manner over a 4-year period with annual measurements. Increases in conscientiousness from year one to year two led to increases in academic behaviors (studying, going to class) from year two to year three, which in turn led to further increases in conscientiousness from year three to year four, which led to even further increases in academic behavior at the terminus of the study at year 4. Essentially, a bidirectional or transactional relationship was in operation between these the two series of variables, making it much less clear which variable was changing the other. Was personality change the leading indicator? Perhaps academic behavior changes in the years before the study had promoted increase in conscientiousness. Yet, perhaps self-control or perseverance increases that date to middle or late childhood preceded those high school academic changes.

The previous example raises the question: which variable was the predictor and which was the outcome? The picture is fuzzy, calling into question our traditional beliefs of what is causal and what is caused. As we have previously noted, there are obviously some outcomes that are true endpoints, such as divorce, disease onset, and mortality (Mroczek & Spiro, 2007; Mroczek, Graham, Turiano, & Oro-Lambo, in press). However, many variables related to personality are not clear points but rather variables that continually change over the lifespan, such as health status and cognitive functioning, which are two variables that form the basis of healthy aging. These variables are likely candidates for entry into bidirectional feedback loops (or multi-directional in the case of three or more variables) with personality, thereby making the predictor-outcome association opaque.

Since Jackson (2011), other studies have provided evidence consistent with personality-related bidirectionality in the areas of physical activity (Allen, Magee, Vella, & Laborde, 2017), work and job experiences (Le, Donnellan, & Conger, 2014), living arrangements (Jonkmann, Thoemmes, Lüdtke, & Trautwein, 2014), relationships (Mund & Neyer, 2014; Robins, Caspi, & Moffitt, 2002) and life events (Kandler, Bleidorn, Riemann, Angleitner, & Spinath, 2012). Further work has shown that changes among traits can be subject to bidirectional processes as well. That is, change in one trait can lead to changes in other traits, feeding back upon the first trait (Klimstra, Bleidorn, Asendorpf, van Aken, & Denissen, 2013). These correlated change or coupled change models (Allemand, Zimprich, & Martin, 2008; Bollen & Curran, 2004), can help establish evidence of bidirectionality, as well as evidence for other relevant concepts such as the "corresponsive principle" (Roberts & Wood, 2006) which holds that people's personality traits lead them to seek out experiences that reinforce those very traits. What this means for healthy aging is that one's state in older adulthood may be the product of longstanding bidirectional processes that in many cases were set in motion decades earlier. The process is continual and mutually reinforcing, and if this promotes optimal or healthy aging, then this is a positive feature of a person's overall health in late life. However, one can envision problematic bidirectional processes, such as high neuroticism setting off frequent fight-or-flight activation that leads to high blood pressure which in turn creates anxiety that maintains or even increases neuroticism. In these cases, it will not necessarily lead to healthy aging.

As we argued elsewhere (Mroczek et al., in press), these evidentiary findings may require us to give up traditional notions of causality. Older ideas (such as those in behaviorism) posited simple stimulus-effect relationships. In contrast, both bidirectionality and "corresponsiveness" (Roberts & Wood, 2006) imply ongoing feedback loops in which cause and effect are difficult or even impossible to disentangle. That said, most bidirectional processes probably have some vague sort of beginning, and thus a distal "cause." Yet once a process is set into motion, feedback loops take over and traditional stimulus-effect causality is very difficult to determine. This is likely how many developmental processes work, including those that define healthy aging. It is also very far away from the simplistic independent variable-dependent variable mindset that still dominates in much of psychology. That said, there may be some processes in healthy aging that look like the lifespan mediation chains that Barker described (long-term chains of linear causality). However, it is likely that more processes in healthy aging are defined by feedback loops in which causality is hard to determine. Healthy aging researchers need to consider these processes, even if they are difficult to model or pin down, because health in late life for any person by definition is something that emerges closer to the end than the beginning of an individual lifespan. Therefore, healthy aging is likely the cumulative product of myriad feedback loops, linear chains, ontogenetic processes, increases or decreases in plasticity, and other developmental phenomena that have brought a person to health and vitality in older age, versus something less than that (in the worst case scenarios, pathological aging).

12.3.2 Differential Lifespan Effects of the Same Trait

In addition to possible feedback loops operating across the lifespan, there is another possibility that may complicate the association between personality and healthy aging. Even if a trait level remains stable, it is possible that the association between that trait and a health variable (and by extension, to healthy aging) varies over different phases in the life course. This interesting possibility was proposed by Shanahan et al. (2014), who argued that the manner in which a given trait influences a health outcome can operate differentially across the lifespan (see also Mroczek, 2014). That is, the mechanisms by which a trait influences health may differ across the lifespan and/or the importance of a trait for health may differ across the lifespan.

For example, Shanahan et al. (2014) suggest that the mechanisms by which self-control influences health differ across the lifespan. In childhood and adolescence, high self-control may be related to the formation of good health habits (choosing better foods, getting exercise). In adolescence and adulthood, high self-control may have its greatest effect through avoidance of risky behaviors such as drug use, risky sex, drunk driving, and thrill-seeking endeavors that can lead to injury, permanent damage to health, or even death. In adulthood, the action mechanism of self-control may center more on medication adherence and engagement with the health system to manage emerging chronic conditions, although diet and exercise likely still play a role here too. Shanahan et al. (2014) also argue that the importance of self-control for health may differ across the lifespan. Specifically, in late adulthood, the effects of self-control may "dissipate" due to biological aging. Such biological processes may overwhelm even the most self-controlled people and halt any effect of self-control on health in this last phase of life. In essence, differential action mechanisms and differential degrees of influence add layers of complexity to any effects of linear chains or of bidirectional feedback loops.

12.3.3 Using Long-Term Linear Chains and Bidirectional Loops to "Target and Tailor"

Regardless of whether a particular personality process leading to healthy (or unhealthy) aging takes the form of a bidirectional feedback loop or a simpler long-term linear chain, personality traits may form the basis for targeting or tailoring efforts that can promote healthy aging as well as other health outcomes (Hagger-Johnson & Whiteman, 2008). The idea behind targeting and tailoring is to use information about individual differences to make interventions more powerful and efficient. It is a form of psychological precision medicine. Targeted and tailored personality interventions are still largely hypothetical. However, we can imagine examples in which they have the potential to be more effective than "one size fits all" approaches. Take our earlier example of neuroticism leading to high blood pressure which reinforces the extant neuroticism. Certain negative affect- and anxiety-reducing interventions, such as meditation or expressive writing, may lower neuroticism and in turn, lower blood pressure. However, these interventions may not work well with people who are extremely high in neuroticism (say, above the 90th percentile). More comprehensive techniques, such as cognitive behavioral therapy (CBT), may work better for these individuals. If so, people who are moderately high in neuroticism may benefit from meditation and expressive writing, whereas people who are extremely high in neuroticism may be better served with a targeted intervention that is based on CBT. In this example, the treatments have been targeted to be more precise and in theory are more powerful because they take into account individual differences.

A successful intervention for these two neuroticism groups – "very high" and "high" – could potentially break the bidirectional coupling of this trait with a physiological variable such as blood pressure. Often, targeted personality-based interventions (which remain largely hypothetical) are seen as valuable because they have the ability to break a linear personality-outcome association. However, their effect may be more far reaching and powerful because they may disrupt a bidirectional feedback loop that has the potential to cycle on indefinitely. "Targeting" refers to when a segment of people is designated for a particular intervention based on an individual difference variable. A slice of the trait spectrum is targeted.

However, "tailoring" often refers to much more intense and specific individual-level intervention. In this sense, tailoring is akin to molecular precision medicine approaches in which a drug or treatment is designed specifically for one individual. Imagine a personality-based tailoring effort whereby an intervention is created for a single person. This is obviously expensive but almost certainly much more powerful than a targeting effort, and probably many times more powerful than the typical "one size fits all" interventions that are common in psychology, medicine, and education.

Personality-based interventions throughout the lifespan have the potential to promote successful aging. Personality-based interventions early in the lifespan can disrupt long-term explanatory chains or bidirectional feedback loops. In older adulthood, personality-based interventions can lead to better proximal outcomes such as higher medication adherence, more physical activity, and greater social engagement. Interventions that are targeted and tailored based on people's individual traits have the potential to be even more impactful than "one size fits all" approaches. However, more research is needed to develop and test such personality-based precision medicine before it can enter the health care toolbox in the coming years.

12.4 Conclusion

This chapter discussed how personality influences heathy aging through distal (long-term) as well as proximal (short-term) predictive chains. Yet we also argued that such linear chains may be too simplistic, and that dynamic processes likely play a major role as well. Specifically, transactional or bidirectional processes in the form of longitudinal feedback loops may describe the manner in which a person ages healthfully or not. We also considered differential predictive power of personality over different life course periods (Mroczek, 2014; Shanahan et al., 2014). Lastly, we discussed a relatively new idea, personality-based interventions, and how they could be part of either targeting or tailoring efforts that promote healthy aging as well as other health outcomes (at any age) by increasing the precision and power of otherwise clumsy, sledgehammer-like interventions (Hagger-Johnson & Whiteman, 2008). Through personality-based precision medicine, lifespan perspectives on personality science could play a key role in promoting the healthful aging of the world's growing population of older adults.

Keeping as much of the world's older adult population as healthy as possible will be a major challenge in the coming decades. Many developed countries that have seen declines in birth rates paired with increases in life expectancy (such as Japan and Switzerland) are already seeing a large increase in the percentage of their populations that are over 60 years of age. Yet the developing world has seen increases in life expectancy as well. In China and India alone, there will be several hundred million new people over the age of 60 by mid-century. Multi-pronged and multidisciplinary efforts will be needed to ensure that as many of these individuals as possible grow older in a healthy manner, retaining functional status and cognitive capacity and maintaining some degree of independence. Insights from personality science could play a key role in those efforts.

References

Aldwin, C. M., Spiro, A., III, & Park, C. L. (2006). Health, behavior, and optimal aging: A life span developmental perspective. In J. E. Birren, R. P. Abeles, & T. A. Salthouse (Eds.), *Handbook of the psychology of aging* (pp. 85–104). Cambridge, MA: Academic.

Allemand, M., Zimprich, D., & Martin, M. (2008). Long-term correlated change in personality traits in old age. *Psychology and Aging, 23*, 545–557.

Allen, M. S., Magee, C. A., Vella, S. A., & Laborde, S. (2017). Bidirectional associations between personality and physical activity in adulthood. *Health Psychology, 36*, 332.

Baltes, P. (1987). Theoretical propositions of life-span developmental psychology: On the dynamics between growth and decline. *Developmental Psychology, 23*(5), 611–626.

Baltes, P., & Baltes, M. (1990). Psychological perspectives on successful aging: The model of selective optimization with compensation. In P. Baltes & M. Baltes (Eds.), *Successful aging: Perspectives from the behavioral sciences* (pp. 1–34). New York, NY: Cambridge University Press.

Barker, D. J. P. (2004). The developmental origins of adult disease. *Journal of the American College of Nutrition, 23*, 588S–595S.

Bollen, K. A., & Curran, P. J. (2004). Autoregressive latent trajectory (ALT) models a synthesis of two traditions. *Sociological Methods & Research, 32*, 336–383.

Friedman, H. S. (2000). Long-term relations of personality and health: Dynamisms, mechanisms, tropisms. *Journal of Personality, 68*, 1089–1107.

Friedman, H. S., Kern, M. L., Hampson, S. E., & Duckworth, A. L. (2014). A new life-span approach to conscientiousness and health: Combining the pieces of the causal puzzle. *Developmental Psychology, 50*, 1377.

Hagger-Johnson, G. E., & Whiteman, M. P. (2008). Personality and health - so what? *The Psychologist, 21*, 594–597.

Hampson, S. E., Edmonds, G. W., Goldberg, L. R., Dubanoski, J. P., & Hillier, T. A. (2015). A life-span behavioral mechanism relating childhood conscientiousness to adult clinical health. *Health Psychology, 34*, 887.

Jackson, J. J. (2011). *The effects of educational experiences on personality trait development.* Doctoral dissertation. Retrieved from ProQuest dissertations & theses global database. (Accession No. AAT 3669204).

Jonkmann, K., Thoemmes, F., Lüdtke, O., & Trautwein, U. (2014). Personality traits and living arrangements in young adulthood: Selection and socialization. *Developmental Psychology, 50*, 683–698.

Kandler, C., Bleidorn, W., Riemann, R., Angleitner, A., & Spinath, F. M. (2012). Life events as environmental states and genetic traits and the role of personality: A longitudinal twin study. *Behavior Genetics, 42*, 57–72.

Klimstra, T. A., Bleidorn, W., Asendorpf, J. B., van Aken, M. A. G., & Denissen, J. J. A. (2013). Correlated change of big five personality traits across the lifespan: A search for determinants. *Journal of Research in Personality, 47*, 768–777.

Le, K., Donnellan, M. B., & Conger, R. (2014). Personality development at work: Workplace conditions, personality changes, and the corresponsive principle. *Journal of Personality, 82*, 44–56.

Mokdad, A. H., Marks, J. S., Stroup, D. F., & Gerberding, J. L. (2004). Actual causes of death in the United States, 2000. *JAMA, 291*(10), 1238–1245.

Mroczek, D. K. (2014). Personality plasticity, healthy aging, and interventions. *Developmental Psychology, 50*, 1470–1474.

Mroczek, D. K., Graham, E. K., Turiano, N. A., & Oro-Lambo, M. O. (in press). Personality development in adulthood and later life. In R. W. Robins, O. P. John, & L. A. Pervin (Eds.), *Handbook of personality: Theory and research* (4th ed.). New York, NY: Guilford.

Mroczek, D. K., & Spiro, A. (2007). Personality change influences mortality in older men. *Psychological Science, 18*, 371–376.

Mund, M., & Neyer, F. J. (2014). Treating personality-relationship transactions with respect: Narrow facets, advanced models, and extended time frames. *Journal of Personality and Social Psychology, 107*, 352–368.

Peel, N. M., McClure, R. J., & Bartlett, H. P. (2005). Behavioral determinants of healthy aging. *American Journal of Preventive Medicine, 28*, 298–304.

Pollack, C. E., Chideya, S., Cubbin, C., Williams, B., Dekker, M., & Braveman, P. (2007). Should health studies measure wealth?: A systematic review. *American Journal of Preventive Medicine, 33*, 250–264.

Roberts, B. W., & Wood, D. (2006). Personality development in the context of the neo-socioanalytic model of personality. In D. K. Mroczek & T. D. Little (Eds.), *Handbook of personality development* (pp. 11–39). Mahwah, NJ: Lawrence Erlbaum Associates.

Robins, R. W., Caspi, A., & Moffitt, T. (2002). It's not just who you're with, it's who you are: Personality and relationship experiences across multiple relationships. *Journal of Personality, 70*, 925–964.

Rowe, J. W., & Kahn, R. L. (1997). Successful aging. *The Gerontologist, 37*, 433–440.

Shanahan, M., Hill, P. L., Roberts, B. W., Eccles, J., & Friedman, H. S. (2014). Conscientiousness, health and aging: The life course personality model. *Developmental Psychology, 50*, 1407–1425.

Smith, T. W. (2006). Personality as risk and resilience in physical health. *Current Directions in Psychological Science, 15*, 227–231.

Turiano, N. A., Chapman, B. P., Gruenewald, T. L., & Mroczek, D. K. (2015). Personality and the leading behavioral contributors of mortality. *Health Psychology, 34*, 51.

World Health Organization. (2017). *10 priorities for a decade of action on healthy ageing*. Retrieved from the World Health Organization website: https://www.who.int/ageing/10-priorities/en/

Chapter 13
Concluding Comments on the Role of Individual Differences in Healthy Aging

Mathias Allemand and Patrick L. Hill

For decades, researchers have emphasized the importance of understanding *how* individual differences predict consequential healthy aging outcomes, such as life quality and longevity. Toward this end, developmental studies have consistently supported the role of personality variables, such as traits, abilities, motives, and goals as informative for promoting health maintenance and improving well-being and quality of life in the context of adult development and aging (Freund, Napolitano, & Rutt, 2019; Hill & Roberts, 2016). This line of research typically uses a structural approach to personality that concerns how the parts of personality are organized together into distinct areas of personality, and what those areas of personality are. It focuses on how persons are different from each other in many ways. A prominent conceptual framework to organize individual differences in relatively stable patterns of experiences and behaviors is the Big Five (John, Naumann, & Soto, 2008) or Five-Factor Model (McCrae and Costa 2008). Research on personality and healthy aging using this structural approach has produced a considerable number of studies that evidenced cross-sectional and longitudinal associations between personality variables and proximal and distal healthy aging outcomes, such as healthy lifestyles and longevity (Friedman & Kern, 2014; Hampson, 2019; Jackson, Weston, & Schultz, 2017). Many contributions in this volume discussed the current literature specifically with respect to personality traits.

M. Allemand (✉)
Department of Psychology and University Research Priority Program Dynamics of Healthy Aging, University of Zurich, Zürich, Switzerland
e-mail: mathias.allemand@uzh.ch

P. L. Hill
Department of Psychological & Brain Sciences, Washington University in St. Louis, Saint Louis, MO, USA
e-mail: patrick.hill@wustl.edu

P. L. Hill, M. Allemand (eds.), *Personality and Healthy Aging in Adulthood*, International Perspectives on Aging 26,
https://doi.org/10.1007/978-3-030-32053-9_13

More recently, researchers have noted the importance of understanding *why* and *when* individual differences have consequential effects on individual functioning in everyday life of adults. The focus of this line of research is on the contextualized dynamic processes as they occur in everyday life (Beckmann & Wood, 2017). A process approach to personality concerns how the different parts of personality influence one another and affect individuals' behaviors and reactions to situations. Research on personality and healthy aging has begun to identify and study momentary and varying personality processes, such as state expressions or regulatory processes, as mechanisms that unfold over short-term time periods to produce the effect of the relatively stable personality variables (e.g., Diehl & Hooker, 2013; Hampson, 2012). As illustrated in this volume, there are currently many research efforts to investigate the role of contextualized dynamic processes underlying healthy aging using ambulatory assessment or mobile sensing approaches.

Almost two decades ago, Hooker and McAdams (2003) proposed a descriptive personality model, called the six foci of personality, to integrate both lines of research. This model integrates structures and processes within one conceptual framework and includes traits, personal action constructs, such as goals, and life stories as structural components, and states, self-regulation, and self-narration as the parallel process constructs. The current volume highlights international efforts to study and understand individual factors involved in healthy aging. It aimed to bring together both lines of research and to present some new findings, developments, and techniques in order to continue progress for research on personality and healthy aging. We have organized the contributions in this volume around three broad sections: constructs, methods, and questions.

13.1 "New" Constructs Beyond the Big Five

As previous research has focused almost exclusively on the Big Five personality traits, one promising avenue to better understand the role of personality in healthy aging is to study constructs beyond the Big Five framework. The chapters in this section made several contributions to the field of personality and healthy aging research. In the initial chapter, Hannah Brazeau and William Chopik discussed how personality and relationship processes may interact to influence physical and mental health in adulthood. In addition to the Big Five traits, they discussed the role of individual differences in interpersonal interactions in terms of attachment processes for health. Attachment is an "old" and established construct in personality and developmental psychology (Fraley, 2019). Despite the fact that attachment processes were not only proposed for childhood but for the entire lifespan, very little is known about how attachment processes manifest in old age and how these processes are associated with daily social functioning and healthy aging processes. The age-related increase in socio-affective orientation in older adults (Carstensen, Isaacowitz, & Charles, 1999) makes attachment processes in old age especially important. Compared to earlier phases of the lifespan, the social world of older adults could be seen as primarily made up of attachment figures, whereas younger adults have a greater social network involving a higher number of peripheral contacts.

Examining the role of personality in healthy aging does not necessarily mean to focus only on positive, healthy, and successful aspects of personality and aging but should also include negative, unhealthy, and less successful individual differences in personality and aging, such as psychopathology. The chapter by Patrick Cruitt elucidates the value of considering personality disorders, in addition to "normal-range" personality, in order to better predict healthy aging outcomes. Though work has only begun to examine personality disorders later in life, the results are quite striking with respect to the added value in predicting important adult developmental outcomes (e.g., Cruitt & Oltmanns, 2018). This research provides a clearly important future direction for research on personality and aging, which again has been somewhat handcuffed by its focus on specific structural approaches to conceptualizing personality.

Gabrielle Pfund and Nathan Lewis highlighted the importance of studying individual differences in purposefulness through a lifespan lens, since sense of purpose seems to be an important contributor and correlate of healthy aging, beginning as early as adolescence. The extent to which people feel that they have personally meaningful goals and directions guiding them through life promotes better psychological, physical, and cognitive health across the entire lifespan (Hill, Edmonds, & Hampson, 2019; Hill & Turiano, 2014). However, further knowledge is needed about how purposefulness promotes healthy aging through different pathways as well as how purposefulness can be imbued via intervention efforts.

The last contribution in this section by Marko Katana and Patrick Hill covered the domain of affective aging. More specifically, the chapter organized the literature on affective experiences along different time-scales and the distinction into personality structures and processes, respectively. Traditional aging research examines how affective traits develop across the adult years into old age. As a complementary perspective, research studying affective experiences in everyday life of older adults focuses more on short-term dynamic processes using state-of-the-art assessment techniques such as ambulatory assessment to measure affective states as they occur in everyday life (Trull & Ebner-Priemer, 2014). This chapter's strong emphasis on assessment methods and designs forms a bridge to the more assessment and analytic-oriented chapters on healthy aging.

13.2 "New" Assessment and Analytical Methods

A second promising avenue to better understand the role of personality in healthy aging is to use assessment and analytical methods beyond the classical approaches to capture long-term developmental processes underlying healthy aging. Integrative data analysis approaches have attracted a lot of attention in the last years (Curran & Hussong, 2009; Hofer & Piccinin, 2009). One of those approaches is coordinated data analysis that refers to the analysis of multiple publicly available datasets in a way to maximize the opportunities for direct comparisons of results (Hofer & Piccinin, 2009). The chapter by Sara Weston and colleagues introduced this approach. It includes five important recommendations derived from a number of conducted coordinated analysis, using a specific empirical example to illustrate this method. Due to the fact that several publicly longitudinal datasets with large samples sizes, broad age ranges, and large time intervals are available, personality and healthy aging researchers should make more use of this unique tool to study how, why, and when individual difference variables, such as traits, abilities, motives, and goals are informative for promoting health maintenance and improving well-being and quality of life across adulthood into old age.

Research interests in dynamic personality and healthy aging processes in everyday life requires adapted research designs and assessment methods that go beyond classical assessment approaches to capture developmental personality change. In their contribution, Joshua Jackson and Emorie Beck discussed new ambulatory assessment approaches that rely on repeated assessments of personality processes where people respond multiple times a day for multiple days. Although ambulatory assessment is a powerful modern methodology that has attracted a great deal of attention in psychology (Trull & Ebner-Priemer, 2014), very few studies have utilized this assessment method as a means to better understand personality and healthy aging processes. The chapter described dynamic metrics to describe personality processes and introduced idiographic approaches to personality that will advance research on personality and healthy aging. Ambulatory assessments typically include momentary self-reports but can also integrate passive assessment methods such as naturalistic observation (e.g., audio or video recording, activity monitoring) and physiological methods (e.g., physiological sensors) (Allemand & Mehl, 2017).

Recent advances in mobile sensing technologies open up new avenues for the use of the smartphone to collect behavioral data (Harari et al., 2016). These technological advances are directly relevant to understanding personality processes and healthy aging. Two contributions in this volume were devoted to innovative assessment methods using mobile sensing. In their chapter, Burcu Demiray and colleagues described the Electronically Activated Recorder (EAR), a naturalistic observation method, which is a portable app-based audio recorder that periodically records sounds and speech in everyday life (Mehl, 2017). Although this assessment method has been used for several years in psychology, only recently has this method been applied to research on healthy aging. The authors argued that EAR data is particularly uniquely valuable when measuring individual differences in social activities and everyday cognitive activities in everyday life. These activities are potential pathways to explain how personality traits influence healthy aging outcomes.

Another assessment method that is underrepresented in the field of personality and healthy aging is the Global Positioning Systems (GPS) to continuously and unobtrusively track locations and mobility. Borrowing from the traditional distinction into personality structures and processes, the chapter by Michelle Fillekes and colleagues distinguished between motility as the relatively stable mobility potential and movement as the manifested mobility. In particular, the chapter presented several indicators of motility and mobility that can be measured with different methods including GPS. These indicators should be tested in future research. Finally, the authors presented conceptual ideas how mobility and personality may promote healthy aging.

13.3 "New" Questions

Recent research in the field of individual differences and healthy aging has accumulated evidence for the importance of personality science in aging processes. However, the current findings have raised new important questions. For example, the fact that personality predicts important healthy aging outcomes leads to the question of whether and how personality can be targeted or modified by psychological and other interventions (Chapman, Hampson, & Clarkin, 2014). The chapter by Damaris Aschwanden and Mathias Allemand introduced a conceptual framework for intervention efforts in which personality plays an important role. The Activities in Motion and in Action (AMA) framework aims to explain how and why engaging in cognitive, physical, and social activities is related to short-term healthy aging outcomes. The authors suggested personality-informed interventions to promote engagement in cognitive, physical and social activities of older adults. The primary idea is with the help of personality-informed interventions to ensure that older adults are "in motion and take action" for healthy and active aging. This AMA framework may stimulate future advances in theory and research.

Several aspects of the AMA model are aligned with recommendations for how to promote healthy cognitive aging. In their contribution, Brennan Payne and Monika Lohani provided an integrative review of the multiple pathways through which personality traits and cognitive functioning interact in promoting healthy aging. Their contribution clearly demonstrated that the field of healthy cognitive aging can benefit from personality science and vice versa (cf. Hill & Payne, 2017). In this respect, the chapter calls for more integrative research efforts to take research on personality and healthy aging to a new dimension. Throughout the contributions in this volume, there was a call for greater effort to integrate more long-term developmental change research with process research on short-term dynamics in everyday life as a necessary step to better understand healthy aging processes.

The last chapter by Daniel Mroczek and colleagues argued from a lifespan development perspective that the "seeds of healthy aging are sown much earlier in life, through the health choices we make in younger adulthood and midlife." Although old age is a unique developmental period with its own challenges and opportunities, research on personality and healthy aging processes should focus on both distal factors, such as midlife precursors of psychological functioning in old age, as well as on more proximal processes, such as how older adults engage in a variety of processes and activities to maintain life quality and health in everyday life (Shanahan, Hill, Roberts, Eccles, & Friedman, 2014). The chapter rightly pointed out that the dynamic processes underlying healthy aging are complex. Existing conceptual models of how personality influences healthy aging processes are often too simplistic. Therefore, future work should think "outside of the box" and develop more comprehensive lifespan models of healthy aging.

13.4 Conclusion

In this volume, we presented advances that scientists have made in recent years to better understand how individual differences predict consequential healthy aging outcomes and why and when individual differences may matter in everyday life. All contributions suggested imperative questions for researchers to pursue in the years to come. The field clearly has a strong foundation of research demonstrating the associations between personality and healthy aging outcomes, but the current volume emphasizes the gaps in the current literature, thus highlighting several new directions, approaches, and considerations for research on personality and healthy aging.

References

Allemand, M., & Mehl, M. R. (2017). Personality assessment in daily life: A roadmap for future personality development research. In J. Specht (Ed.), *Personality development across the lifespan* (pp. 437–454). San Diego, CA: Elsevier.

Beckmann, N., & Wood, R. E. (2017). Editorial: Dynamic personality science. Integrating between-person stability and within-person change. *Frontiers in Psychology, 8*, 1486.

Carstensen, L. L., Isaacowitz, D. M., & Charles, S. T. (1999). Taking time seriously: A theory of socioemotional selectivity. *American Psychologist, 54*, 165–181.

Chapman, B. P., Hampson, S., & Clarkin, J. (2014). Personality-informed interventions for healthy aging: Conclusions from a National Institute on Aging work group. *Developmental Psychology, 50*, 1426–1441.

Cruitt, P. J., & Oltmanns, T. F. (2018). Incremental validity of self- and informant report of personality disorders. *Assessment, 25*, 324–335.

Curran, P. J., & Hussong, A. M. (2009). Integrative data analysis: The simultaneous analysis of multiple data sets. *Psychological Methods, 14*, 81–100.

Diehl, M., & Hooker, K. (2013). Introduction to a special issue: Adult personality development: Dynamics and processes. *Research in Human Development, 10*, 1–8.

Fraley, R. C. (2019). Attachment in adulthood: Recent developments, emerging debates, and future directions. *Annual Review of Psychology, 70*, 401–422.

Freund, A. F., Napolitano, C. M., & Rutt, J. L. (2019). Personality development in adulthood: A goal perspective. In D. P. McAdams, R. L. Shiner, & J. L. Tackett (Eds.), *Handbook of personality development* (pp. 313–327). New York, NY: Guilford.

Friedman, H. S., & Kern, M. L. (2014). Personality, well-being, and health. *Annual Review of Psychology, 65*, 719–742.

Hampson, S. E. (2012). Personality processes: Mechanisms by which personality traits "get outside the skin". *Annual Review of Psychology, 63*, 315–339.

Hampson, S. E. (2019). Personality development and health. In D. P. McAdams, R. L. Shiner, & J. L. Tackett (Eds.), *Handbook of personality development* (pp. 489–502). New York, NY: Guilford.

Harari, G. M., Lane, N. D., Wang, R., Crosier, B. S., Campbell, A. T., & Gosling, S. D. (2016). Using smartphone for collecting behavioral data in psychological science: Opportunities, practical considerations, and challenges. *Perspectives on Psychological Science, 11*, 838–854.

Hill, P. L., Edmonds, G. W., & Hampson, S. E. (2019). A purposeful lifestyle is a healthful lifestyle: Linking sense of purpose to self-rated health through multiple health behaviors. *Journal of Health Psychology, 24*(10), 1392–1400.

Hill, P. L., & Payne, B. R. (2017). Don't forget the person when promoting healthy cognitive aging: A comment on smith (2016). *American Psychologist, 72*, 390–392.

Hill, P. L., & Roberts, B. W. (2016). Personality and health: Reviewing recent research and setting a directive for the future. In K. W. Schaie & S. L. Willis (Eds.), *Handbook of the psychology of aging* (8th ed., pp. 205–218). Amsterdam, The Netherlands: Elsevier.

Hill, P. L., & Turiano, N. A. (2014). Purpose in life as a predictor of mortality across adulthood. *Psychological Science, 25*, 1482–1486.

Hofer, S. M., & Piccinin, A. M. (2009). Integrative data analysis through coordination of measurement and analysis protocol across independent longitudinal studies. *Psychological Methods, 14*, 150–164.

Hooker, K., & McAdams, D. P. (2003). Personality reconsidered: A new agenda for aging research. *The Journals of Gerontology. Series B, Psychological Sciences and Social Sciences, 58*, 296–304.

Jackson, J. J., Weston, S. J., & Schultz, L. H. (2017). Personality development and health. In J. Specht (Ed.), *Personality development across the lifespan* (pp. 371–384). San Diego, CA: Elsevier.

John, O. P., Naumann, L. P., & Soto, C. J. (2008). Paradigm shift to the integrative big five trait taxonomy: History, measurement, and conceptual issues. In O. P. John, R. W. Robins, & L. A. Pervin (Eds.), *Handbook of personality: Theory and research* (3rd ed., pp. 114–158). New York, NY: Guilford.

McCrae, R. R., & Costa, P. T., Jr. (2008). The five-factor theory of personality. In O. P. John, R. W. Robins, & L. A. Pervin (Eds.), *Handbook of personality: Theory and research* (3rd ed., pp. 159–181). New York, NY: Guilford.

Mehl, M. R. (2017). The electronically activated recorder (EAR): A method for the naturalistic observation of daily social behavior. *Current Directions in Psychology, 26*, 184–190.

Shanahan, M. J., Hill, P. L., Roberts, B. W., Eccles, J., & Friedman, H. S. (2014). Conscientiousness, health, and aging: The life course of personality model. *Developmental Psychology, 50*, 1407–1425.

Trull, T. J., & Ebner-Priemer, U. (2014). The role of ambulatory assessment in psychological science. *Current Directions in Psychological Science, 23*, 466–470.

Printed in the United States
by Baker & Taylor Publisher Services